HPBooks

WELDER'S HANDBOOK

Richard Finch
&
Tom Monroe, P.E.

ANOTHER FACT-FILLED AUTOMOTIVE BOOK FROM HPBooks

Publisher: Rick Bailey; Executive Editor: Randy Summerlin; Editorial Director: Tom Monroe, P.E., S.A.E.; Art Director: Don Burton; Book Design: Paul Fitzgerald; Managing Editor: Cindy Coatsworth; Typography: Michelle Carter, Beverly Fine; Director of Manufacturing: Anthony B. Narducci; Photos: Richard Finch, others noted; Cover photo: Bill Keller

NOTICE: The information in this book is true and complete to the best of our knowledge. All recommendations are made without guarantees on the part of the authors or HPBooks. The authors and publisher disclaim all liability in connection with the use of this information.

Published by HPBooks, a division of HPBooks, Inc.
P.O. Box 5367, Tucson, AZ 85703   602/888-2150
© 1985 HPBooks, Inc.   Printed in U.S.A.
3rd Printing

Library of Congress Cataloging-in-Publication Data

Finch, Richard.
  Welder's handbook.

  Includes index.
  1. Welding—Handbooks, manuals, etc.   I. Monroe,
Tom, 1940-   . II. Title.
TS227.F5   1985   671.5'2   85-60515
ISBN 0-89586-257-3

CONTENTS

ACKNOWLEDGMENTS

I couldn't have written this book without the help of my wife, Breven. She did the typing, ran most of the errands, and even let me teach her how to arc weld, gas weld and solder so I could prove my teaching methods. She also took the photographs as I welded, and followed me to welding shops, taking notes as I took pictures.

Bert Snedden of Hopper Steel in Santa Maria, California helped with welding supplies and photo sessions at their welding/steel-supply store. Ralph Payne, aircraft designer and fabricator, Charles and Louis Hoffman, professional weldors, and Edwin Monroe of Cameron Tool & Supply, provided technical information.

Thanks to various vendors listed throughout this book, but especially Linde Welding & Cutting Systems, Victor Equipment, K. Woods, Inc., Miller Electric Manufacturing Co., Hobart Brothers Welders, and The James F. Lincoln Arc Welding Foundation.

Special thanks to Allan Hancock College in Santa Maria, California and to Ron Smith, President of Aerostar Corporation. Thanks also to Dennis Elias of DE Engineering & Fabrication and Don Bevers of Bevers Enterprises, both of Tucson, Arizona. Dennis and Don opened up their shops so welder and equipment photos could be shot.

Finally, there are the editors at HPBooks, especially Ron Sessions, who made the manuscript readable, added a lot of information and asked me hundreds of questions. Without them, this book would not have been possible.

Introduction

I once saw a bumper sticker that read . . . "THE WORLD IS HELD TOGETHER BY WELDS." Just imagine the consequences if everything that was welded suddenly became "unwelded." Cars would fall into little pieces. Airplanes would come apart and fall from the sky. Your stove or refrigerator would clatter to the kitchen floor. Most public buildings, bridges and the like would crumble as their super-structures came apart. Welds are very important. They really do hold together many man-made things.

Because so many things we are interested in are held together by welds—race cars, airplanes, farm machinery, ornamental ironwork and the like—many people would like to weld or already do. The problem is the ones that don't weld think it's too hard. Not so. Welding is similar to playing a musical instrument. You must first have some talent capable of being developed. With a little practice, you may be able to play Chopsticks on the piano.

Although playing the piano with great skill is something most of us will never do, there's a level of skill and competence in between. Similarly, you should be able to run a weld bead with a little practice. There is a usable level of skill that most people can master in a short period of time. However, the more you practice, the better your welding skills should become. For example, I've taught people how to gas-weld 4130-steel aircraft tubing in about one hour. I've even taught beginners how to pass aircraft-welding certification tests in less than 40 hours. Of course, I can't teach each of you in person. But the next best thing is to read this book, follow my suggestions, and learn at your own pace!

I begin by describing the basics of welding. Then, I briefly describe

THE AUTHORS

Richard Finch, above, and the experimental Aerostar 602P which he flew as the test pilot. When a Power Plant Engineer for Piper Aircraft Corporation, Richard designed the aluminum intake and sonic nozzle system for its twin-turbocharged Lycoming engines. He also welded these components, the engine mounts, stainless-steel exhausts and various other parts for the experimental 602P.

Richard learned welding the hard way—by trial and error. He began welding on drag-race cars, farm tractors and trailers while in high school and college. His first significant welding started with building go-karts. This progressed to building road-racing cars and experimental aircraft. During this time, he became a certified aircraft weldor and a certified nuclear power-plant welding inspector.

In his "spare time" Richard earned two B.S. degrees in engineering, an Associate of Science degree in welding and taught welding at Allan Hancock College in Santa Maria, CA.

Tom Monroe started welding while in high school. It helped pay his way through college. After graduation, Tom took a position with Ford Motor Company as a chassis-design engineer. One of his assignments involved the design and testing of LeMans, CanAm and TransAm race cars at Kar Kraft—Ford's captive race group.

Later, Tom launched his own engineering-consulting business. Activities included the design, build and testing of Pro-Stock drag-race cars, late-model stock cars, IMSA GT and RS cars, NASCAR Modifieds, Formula Super V's and off-road-race cars. During the mid-70s, he was also the chief engineer on the Bricklin sports car. He is now the Automotive Editorial Director at HPBooks.

Above photo by Bob Davis.

each type of welder so you can choose which is best for you. Because gas welding is basic to most types of welding, I concentrate on it, describing in detail how to gas-weld various joints and materials. So, learn how to gas weld *before* you attempt another type of welding. You'll be a better *weldor* because of it. After discussing gas welding, I continue by describing popular welding techniques, equipment required to do each, and which materials can be welded.

Weldor or Welder?—I use the terms *weldor* and *welder*. This is to avoid confusion. Weldor refers to the person; welder refers to the machine. Let's now look at how the weldor uses the welder!

T-bucket project utilized three different welding techniques by the builder, Larry Hofer. Frame was arc welded, exhaust system oxyacetylene welded, and aluminum body TIG welded. Photo by Tom Monroe.

So, you want to learn to weld! You *can* do it. But, before you start, there are some basics you should know. Let's begin with a short course in physics.

Most matter is in one of three basic forms: solid, liquid or gas. We know that at 70F (21C) and normal atmospheric pressure, steel is a solid, water is a liquid and oxygen is a gas. However, matter can be converted from one state to another by raising or lowering its temperature. For example, water freezes and becomes a solid below 32F (0C). Heat it to boiling and it changes to a gas—212F (100C) at sea-level pressure. Metals can also change to a liquid or gas when heated to extreme temperatures.

Fusion Welding—In this book, I cover *fusion welding* as well as brazing and soldering. Regardless of the energy, or heat, source—gas flame, electric arc or electric resistance—fusion welding is the act of heating metal to a liquid state at the weld seam, *fusing* it to (combining it with) another metal and allowing the seam to cool and return to its solid state. The additional metal is usually a *filler metal*—welding rod, sometimes called an *electrode* when used for arc welding—that forms the weld bead. In rare cases, such as when welding exotic metals, you can actually cut a thin strip of the metal you're welding and use it as the welding *rod*.

Gas Welding—The basic and most common welding technique is gas welding. In this process, two gases are used: the fuel and oxygen. Consequently, this type of gas welding is sometimes referred to as *oxyfuel gas welding*. The fuel—acetylene, propane, natural gas, methylacetylene propadiene (MPS), propylene or hydrogen—is mixed with oxygen and ignited at the *torch* to produce the needed heat energy to raise the work and filler metal to their melting temperatures so they can be fused together.

Electric Welding—Most types of electric welding use an electric arc to generate the needed heat energy for the welding operation. The common types include arc welding, *tungsten inert-gas* (TIG)

Regardless of the degree of sophistication, all fusion-welding methods use the same basics. MIG welder is being used to weld frame subassembly that's held in a fixture. Photo by Tom Monroe.

welding and *metal inert-gas* (MIG) welding. TIG welding is also called *Heliarc* welding, a Linde trade name. A common name for MIG welding is *wire-feed welding* because of the way the filler material is *fed* into the weld bead. The small-diameter welding *wire* is fed continuously from a spool to the weld seam.

Two major ways these electric-welding processes differ from each other is how the weld puddle is *shielded,* or protected, from atmospheric contamination, and the type of electrode used.

Shielding is usually done with inert gas or with a *coating* on the filler rod. Arc welding uses coated rods, and TIG and MIG welders usually use gas shielding. When the coating—*flux*—on the arc-welding rod melts with the rod, it generates a gas and floats to the surface of the weld puddle, shielding and forming an insulating coating called *slag* on the weld bead. This slag coating can be *chipped* off, or in the case of a good weld made with *low-hydrogen* stick elec-

trode, peels itself off as it cools.

The gas shielding of a TIG or MIG welder is directed around the electrode at the weld puddle. No slag is formed on either weld.

Electrodes used with arc and MIG welders are the same type. Each is *consumable*—the electrode also acts as the filler material. It is consumed, or melted, to form the weld puddle. TIG-welding electrodes are *non-consumable.* An electric arc is maintained between a tungsten tip and the work to generate heat energy. The filler material is introduced into the weld puddle separately, as it would be in gas welding. Although tungsten electrode wears down and must be *dressed,* or sharpened on occasion, it is not consumable.

I also cover *plasma-arc* welding, a process similar to TIG welding.

Brazing & Soldering—*Brazing* and *soldering* are two gas-welding processes that join base metals by melting a welding rod that builds up thickness or *fills* a weld joint to *bond* the base metal as it cools. Unlike welding, the base metal is *not melted* and fused together when brazing or soldering. Lower temperatures are involved, too.

Flux—A cleaning agent called *flux* is sometimes required for certain types of welding such as brazing, soldering or, as previously mentioned, arc welding. Flux removes impurities from the base-metal surface and sometimes insulates the weld bead for controlled cooling. It melts when heated. Flux is in powder or paste form for brazing, and soldering, or as a hard coating on arc-welding and some brazing rods.

Without flux, the filler material would not adhere to the *base metal*—metal being welded—or, in case of arc welding, the weld bead would oxidize. In both cases, the weld joint would be weak.

Powder or paste-type flux is usu-

ally applied to the filler rod. But, it can also be applied to the joint being soldered or brazed. The weldor does this by heating the end of the rod and plunging it in the flux. This is done as often as necessary to keep the rod coated as it is consumed during welding.

Temperature Control—If you can learn to correctly control the temperature of the *weld puddle*—the molten pool that forms the weld bead—you can do a good job of welding. To control the weld puddle, you must learn how to judge and *control* the temperature of the metal you're welding.

Getting back to the water example, water will pour and flow at room temperature. However, if you lower its *temperature* to below its *melting point,* or 32F (0C), it becomes a *solid*—ice. You can then handle it just like a block of steel; saw it, drill it or sit it on the freezer shelf without the need to contain it.

If you put this block of solid water in a pan and heat it to 212F (100C), it begins to boil and vaporize as it changes from a solid, to a liquid, then to a gas. Steam boils off and escapes into the atmosphere. Water, therefore, has a freezing point, a melting point and a *boiling point.*

WELDING TEMPERATURE

With the frozen state of water in mind, consider the same condition for a piece of steel plate. When the steel is "frozen," below about 2700F, its melting point, it is solid. When heated above that temperature, steel changes to a liquid. And if you heat it to a temperature considerably above the melting point, steel can boil and vaporize into the atmosphere. You obviously shouldn't *vaporize* a weld bead, so keep the weld-puddle temperature below the boiling point.

Frozen water—ice—is *solid* below its 32F (0C) melting point. Steel is also solid below its melting point of 2700F (1482C).

As water temperature rises, a *molten* puddle—liquid water—forms. Heat steel above its melting point, and it too will form a molten puddle.

Additional heat applied to molten puddle causes water to boil and *vaporize*—change to a gas. The same thing happens to steel if you apply excess heat from 6300F (3482C) flame of oxyacetylene torch.

COLOR OF STEEL AT VARIOUS TEMPERATURES in Fahrenheit (Centigrade)

Faint Red	900 (482)
Blood Red	1050 (566)
Dark Cherry Red	1075 (579)
Medium Cherry Red	1250 (677)
Cherry Red	1375 (746)
Bright Red and Scaling	1550 (843)
Salmon and Scaling	1650 (899)
Orange	1725 (941)
Lemon	1825 (996)
Light Yellow	1975 (1079)
White	2200 (1204)
Dazzling White	2350 (1288)

Colors are as viewed in medium light, not in bright sunlight or total darkness.

COLORS FOR TEMPERING CARBON STEEL

Color	Metal Temp F (C)
Pale Yellow	428 (220)
Straw	446 (230)
Golden Yellow	469 (243)
Brown	491 (255)
Brown and Purple	509 (265)
Purple	531 (277)
Dark Blue	550 (287)
Bright Blue	567 (297)
Pale Blue	610 (321)

See Page 64 for information on how to temper screwdriver blades and other carbon steels.

Carbon steel assumes a color when heated. It is then cooled to *temper*, or increase hardness.

Most beginning weldors don't heat the base metal enough. They fail to get it to the melting point and keep it there. I usually tell the beginner to go ahead and *burn up*, or overheat a few practice pieces. Doing this gives the beginner a feel for *temperature control*. Melting temperatures vary considerably from metal to metal. If you can master temperature control, you have a head start on welding successfully.

Once you've mastered temperature control, the next thing to do is to become familiar with which metals can be welded and by which methods. The accompanying charts give the melting points of metals that can be welded, brazed, and soldered.

Make photocopies—you have the author's and publisher's permission—of the five charts in this chapter. Post them in your welding shop for quick reference. You'll be amazed how helpful they'll be.

This chapter gives you the basics of soldering, brazing and welding. Read this chapter and the gas-welding chapter, page 45, to become familiar with welding technique *before trying any kind of welding*—gas, arc, MIG, TIG, plasma or whatever.

IDENTIFYING TYPES & KINDS OF METALS

Before you can begin welding, you must know what metal you're going to weld. For instance, it's apparent that you can't weld this page to your car's radiator. But other combinations may not be so obvious.

I've often been asked to weld metals that are not weldable. And I have received similar requests to join metals that cannot be soldered or brazed. Perhaps it's welding dissimilar metals together such as aluminum to steel, or steel to pot metal. I've even been asked to weld 2024 aluminum; the experts

WEIGHTS, MELTING POINTS & BOILING POINTS OF METALS

Metal	Wt. Per Cu Ft in Lb	Melting Point in Degrees F (C)	Boiling Point in Degrees F (C)
Aluminum	166	1217 (658)	4442 (2450)
Bronze	548	1562—1832 (850—1000)	
Brass	527	1652—1724 (900—940)	
Carbon	219	6512 (3600)	
Chromium	431	3034 (1615)	
Copper	555	1981 (1083)	4703 (2595)
Gold	1205	1946 (1063)	5380 (2971)
Iron	490	2786 (1530)	5430 (2999)
Lead	708	621 (327)	3137 (1725)
Magnesium	109	1100 (593)	
Manganese	463	2300 (1260)	
Mild Steel	490	2462—2786 (1350—1530)	
Nickel	555	2645 (1452)	4950 (2732)
Silver	655	1761 (960)	4010 (2210)
Tin	455	449 (231)	4120 (2271)
Titanium	218	3263 (1795)	
Tungsten	1186	5432 (3000)	10706 (5930)
Zinc	443	786 (419)	1663 (906)

Melting points of metals vary widely. If available, I've also included each metal's boiling point, a temperature you should avoid.

TEMPERATURES OF SOLDERING, BRAZING & WELDING PROCESSES

Soldering, Lead Solder	250—800F (121—427C)
Brazing, Brass and Bronze	800—1200F (427—649C)
TIG Welding	5000F (2760C) Arc-Temp Variable
Oxyacetylene Welding	6300F (3482C) Flame Temp
Oxyacetylene Cutting	6300F (3482C) Flame Temp
Arc Welding	6000—10,000F (3316—5538C) Arc Temp
Plasma-Arc Cutting	50,000F (27,760C) Arc Temp

Keep these working temperatures in mind. Each process—welding, brazing or soldering—is different. Overheat the weld bead and you'll "vaporize" your project! Master temperature control and you'll become a much better weldor.

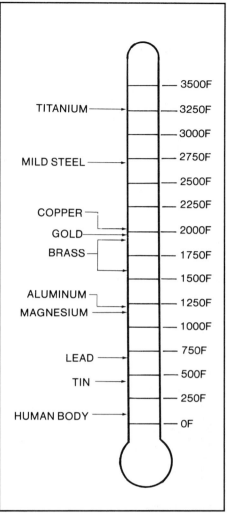

"Thermometer" will help you visualize melting points of various metals. It will also give you an idea of how much heat you'll need for welding various metals.

say this aluminum alloy is not weldable. Many other aluminum alloys are weldable.

Alloy—An alloy of a metal, such as aluminum, steel or titanium, is the basic metal combined, or *alloyed,* with another element(s) to obtain or enhance certain characteristics. Such characteristics include strength, formability and corrosion resistance. The alloying of a metal may also affect its weldability. In the following section, I cover many of the common metals and their alloys that can or cannot be welded.

Ferrous or Non-Ferrous?—I keep a small magnetic screwdriver in my shirt pocket for quick identification of metals. Why? Because I cannot always visually identify the metal I'm working with. As you probably know, a magnet attracts *ferrous* metals—those containing iron. Magnets are not attracted to *non-ferrous* metals—metals that contain little or no iron, such as aluminum.

So, I touch my magnetic pocket screwdriver to the metal I want to weld. If it sticks, the metal has iron content. If the magnet doesn't stick, the metal has little or no iron content. But, this is only a preliminary test. Some non-magnetic metals, such as stainless steel, cannot be fusion-welded to other non-magnetic metals, such as aluminum or magnesium. Such dissimilar metals must be joined by brazing or soldering.

Of course, there's more to identifying metals than determining if they're ferrous or non-ferrous.

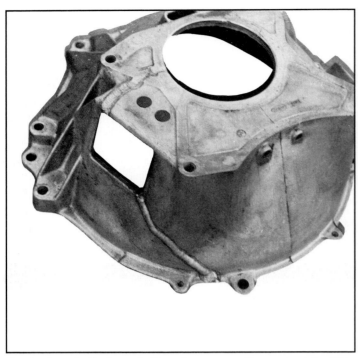

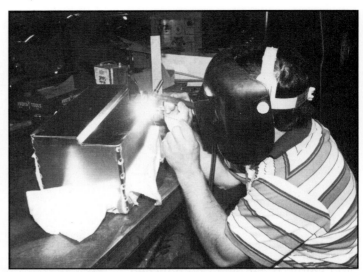

Ron Fournier, author of HPBooks' *Metal Fabricator's Handbook*, TIG-welds aluminum oil tank. Note protective paper covering peeled back from weld area. Paper reduces surface oxidation and protects aluminum from scratches. Photo by Ron Fournier.

Cracked cast-aluminum bellhousing was easily repaired by TIG welding. Because cast aluminum is so easy to weld, high-performance engine builders prefer using aluminum cylinders heads instead of cast iron. Not only are they easier to repair, they are also easier to modify. Photo by Tom Monroe.

Even though cast iron and mild steel are both magnetic and ferrous, the two cannot be joined using conventional welding methods. Special care must be taken to choose the proper welding process and welding rod.

I divide metals into like groups and identify each by simple, visual methods. I begin with the most familiar and progress through the newest, most-exotic metals.

IDENTIFYING METALS

Cast iron is usually rough-textured because of how it is manufactured. A cast-iron part is formed by pouring molten iron into a sand mold, thus giving it the form and texture of the interior of the mold. Typical cast-iron parts include automotive engine blocks, exhaust manifolds, manual-transmission cases, older lawn-mower and garden-tractor engines and early-style farm equipment.

Where cut with a lathe, saw, grinder or whatever, cast iron usually has a gray, grainy appearance. When ground with a high-speed grinding wheel, red sparks are generated.

Cast iron is ferrous, or magnetic. It can be arc-welded with stick-type electrodes or brazed or fusion-welded with an oxyacetylene torch.

Cast brass is often rough-textured because it, too, is usually cast in sand molds. Cast brass is used for water-valve housings and, with bronze alloy, for boat propellers. If you cut or machine cast brass or *bronze*—a brass alloy—it appears smooth and yellow or gold in color, depending on the specific alloy. Brass should not be ground. Soft metals such as brass, *load up*—clog—abrasive grinding wheels, quickly rendering them useless.

Cast brass is non-ferrous and non-magnetic. It can be welded with flux-coated brass rod and an oxyacetylene torch.

Mild Steel, most common of all metals, is used in automobile bodies and chassis parts, lawn-mower handles, bicycle fenders, house furniture, file cabinets—the list is endless. Cut or ground, it looks bright gray to silver. Grinding with a stone wheel generates a shower of yellow sparks.

Mild steel is ferrous. It can be welded by every method described in this book: gas, arc, TIG, MIG, plasma-arc and spot welding. It is very easy to work with.

Forged Steel is rough, but smoother than cast iron. It's used for most engine connecting rods, some crankshafts, axle shafts and some chassis components. Forging steel is done by hammering a red-hot steel billet into the desired shape in a forging press. Machined, cut or ground, forged steel is light gray or silver inside. Grinding forged steel creates yellow or white sparks.

Forged steel is a ferrous metal. It can be welded with gas, arc, TIG, MIG or plasma-arc methods. But, because steel forgings are intended for high-load/high-fatigue applications, such parts should be welded using the best method available—TIG, plasma or d-c arc—if welded at all. Usually, a damaged forged-steel part should be replaced rather than welded.

Cast Aluminum is usually rough, but because it's cast at a much lower temperature than cast iron, it can be cast in molds with a smoother finish. Cast aluminum is used for late-style lawn-mower engines, motorcycle crankcases, intake manifolds for automobile engines and, more recently, automobile-engine blocks and cylinder heads.

As with brass, don't grind aluminum with a stone wheel. It also clogs the abrasive. Cast aluminum is non-ferrous. It can be welded with TIG, plasma and gas, and brazed with aluminum filler. Cast aluminum can also be arc-welded.

Sheet Aluminum is smooth and shiny. It can even be polished to a mirror finish. Sheet aluminum comes in thicknesses as thin as kitchen foil, or as thick as 2- or 3-in. plates. Sheet aluminum is used for screen-door frames, airplane wings, race-car chassis, lawn furniture, siding on buildings and many other common applications.

Sheet aluminum is non-ferrous. Certain alloys can be welded with gas or TIG. You can also arc-weld some sheet aluminum.

Stainless Steel is very smooth and hard. It's usually found in sheets, but can be cast. As the name implies, it will not stain or corrode easily. Stainless steel is used to make kitchen cutlery, pots and pans, exhaust systems for airplanes or autos and, in the early '80s, the DeLorean sports-car outer skin. When ground with a high-speed abrasive stone, it does not give off sparks. Instead, stainless steel turns black where ground.

Stainless steel is an exception to the ferrous rule. It is steel, but it is *usually* non-magnetic. However, a few of the hundred or so different stainless-steel alloys are magnetic. Stainless steel welds best with TIG, but arc and gas welding can be used as well. It can also be gas-brazed.

Titanium looks similar to stainless steel, but it's much shinier when welded or filed. Although relatively light, it is very strong. Titanium-alloy forgings, tubes and sheets are used in aircraft and race-

Stainless-steel headers are superior to mild-steel headers, particularly for high-heat applications such as these for Bob Sharp's turbocharged GT-1 Datsun. Photo by Ron Fournier.

Small, magnetic pocket screwdriver is handy to test for ferrous and non-ferrous metals. In this picture, magnet will not pick up brass welding rod because brass is non-ferrous, thus non-magnetic.

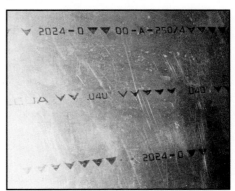

Sheet aluminum is marked for easy identification. This 2024-0 aluminum sheet can be welded.

car construction. It is very expensive.

Titanium doesn't give off sparks when ground with an abrasive wheel. A ferrous metal, titanium can be TIG-welded, but shielding is critical.

VISUAL MARKINGS

Most metal alloys have identification markings:

Chrome-moly steel, for instance, is usually marked 4130 Cond. N, for 4130 steel, *normalized condition.* The condition of metal indicates how its *temper,* or hardness, was achieved. For instance, it may be in an as-fabricated condition, work-hardened or heat-treated. The same applies to tubing and sheet chrome-moly steel.

Sheet aluminum is usually marked to indicate the basic alloy and its conditions. Typical examples of aluminum alloys are 2024-T4, 3003-H14 and 6061-0. Refer to the chart of aluminum alloys, page 102, for a more detailed explanation.

Sheet stainless steel is marked, 301, 308 or 347, for example, depending on the specific alloy. The markings are similar to those for sheet aluminum except that STAINLESS may also be printed on the sheet.

For an in-depth discussion of popular steel and aluminum alloys, read HPBooks' *Metal Fabricator's Handbook.* Not only are important properties of metals discussed, but the specifics of each of the popular alloys are covered.

Safety, Comfort & Convenience

2

Your welding shop should be neat and tidy. Not only will you make better welds, it'll be safer. Note clean floor, fire extinguisher on back wall, large steel table with vise and good lighting. Photo by Ron Fournier.

Welding is more hazardous than most other shop processes. The dangers of fire and explosion, burned hands or metal in the eyes are always present. Additionally, high-pressure oxygen, acetylene, argon, CO_2 or helium tanks present a potential work hazard. Hot metal is always a danger if it comes in contact with anything flammable or meltable, or your skin.

To start with, I'll relate some horror stories—welding explosions caused by carelessness.

First, an explosion that caused a fatality occurred several years ago when a local weldor was assigned the job of cutting several 55-gallon oil drums in half. These were to be made into cattle-feeding troughs. He had cut several empty drums in two, splitting them lengthwise with his cutting torch. On the last drum he cut open, he was straddling it as if riding a horse. Just as his cutting torch pierced the metal, the drum exploded, blowing the weldor through the shop's corrugated roof. It was estimated that he was blown 50-feet straight up.

I don't know what the oil drums contained, but the one that exploded obviously contained a fatal combination of explosive vapors.

An accident that could've been caused by welding was nearly fatal to two small boys. A contractor had been coating asphalt driveways in my neighborhood with *water-soluble* black sealer. At the end of a long, hot workday, the contrac-tor left an almost empty 55-gallon drum of the material sitting beside the driveway of the last home. The two children began throwing lighted matches into the *bung* in the top of the drum. Each match ignited vapors, exploding with a loud ka-whoom. When they threw in the last match, vapors inside the drum exploded and blew the drum about 200 feet into the air. It landed several houses down the street in another yard.

The tar-like substance splattered all over the two boys, but luckily they weren't hurt. If they had been leaning over the drum, they could have been decapitated when the drum took off like a rocket.

Then there was the story one of my auto-shop students related to me about an explosion.

Don't cut or weld a container unless it has been completely purged of all vapors. Even so-called non-flammables can generate explosive vapors when conditions are right. The weldor who blew himself up was building cattle feed troughs such as this. Two 55-gallon oil drums are cut in half and welded together.

This accident occurred in the driveway of a private residence near my house. The destruction was so great that the police and fire departments blocked off three streets to preserve order while they surveyed the damage.

A local oil-trucking company had hired two neighborhood weldors to weld taillight brackets on the rear bumper of some new trucks. The trucks were really fancy, with fiberglass cabs, chrome-plated dual exhaust stacks, and a large stainless-steel tank on the back for hauling crude oil. The tank was a two-layer assembly with fiberglass insulation between the inner tank and the outer cover. The welding had been completed on one truck. The second truck was backed into the driveway so the weldor could get the welding cables to the rear bumper. The truck was so big that it occupied the entire driveway.

The auto-shop student had been watching, but went home for lunch. He lived in a house one block from where the welding was being done. He heard a tremendous explosion and ran outside just in time to see a huge stainless-steel tank fall on top of the house behind his. At first he thought a missile from nearby Vandenberg AFB had exploded and was falling on the neighborhood. Then he realized the debris had come from where the trucks were being welded, so he ran down the street to find *the new truck had been totally demolished!* One of the weldors had been blown across the street, bruised but alive. The other fellow was found inside the kitchen of the house, also bruised but alive.

The truck didn't fare as well. The oil tank, of course, had been blown off the frame, straight up 100 feet or more and had parked itself on the house roof. The cab

was laying shattered in the next yard. The two chrome-plated exhaust pipes had been flattened to the ground. The frame of the truck was warped and twisted. Pieces of the truck landed on cars and the rooftops of several nearby houses. The new truck was totaled! Damage to the house was 50%. The two men welding on the truck miraculously escaped with only minor injuries. Hello, Prudential?

The explosion was caused by a welding spark igniting vapors escaping from the crude-oil tank.

The lesson to be learned from these accidents is that containers of flammable products should not be welded on or cut with a torch. It is safest to refuse to weld on or near any such tank or container. Even vapors from *non-flammable* liquids can be *explosive* under certain conditions. And for sure, vapors from flammable liquids are explosive.

As a weldor, people will ask you to weld many different types of containers. You may even want to weld something similar yourself, but the safe thing is to not do it.

If you *must* have a tank welded, take it to a professional welding shop that specializes in welding gas, oil and similar flammable tanks. They'll know what preparation and welding procedures are needed to prevent accidents such as just described. Typically, a professional welding shop will "boil out" the tank in radiator cleaning acid to remove any oil or gas residue, then purge the tank with an inert gas such as argon while welding it.

ARC-WELDING SAFETY
Radiation Burn—The primary safety hazard in arc-welding is from ultraviolet-light (radiation) burns to the eyes and, to a lesser degree, the skin. This hazard ap-

Protect yourself from ultraviolet (UV) rays and sparks while welding. My wife is wearing denim jeans, long-sleeve sweat shirt, gloves and a good helmet. She is also wearing lace-up shoes. Avoid double-knit or polyester clothing when welding.

I like this arc-welding helmet because it weighs less than 1 lb. Made of treated fiberboard, it's much lighter than a fiberglass helmet.

Padded headband provides comfort and keeps sweat from running down your forehead and into the eyes. Knob adjusts headband for different head sizes.

Don't buy a helmet with pin-and-hole headband adjustment such as this. I could never get it to fit. As a result, helmet wouldn't stay in place. I keep this one around for friends who want to watch me weld.

plies to TIG, plasma-arc and wire-feed welding as well.

The burns received from electric-arc welding are similar to sunburn, except usually deeper into the skin or eyes. One reason for the deeper and potentially more severe burns may be the ultraviolet light source—it's much closer to the body than the sun is. Therefore, arc-welding radiation is more intense. And there's less atmospheric dust to filter the rays of the arc welder.

Clothing—The solution for preventing burns is simply to shield the skin and eyes from ultraviolet light. You should wear a long-sleeve, close-weave shirt and trousers of material least likely to ignite from sparks. Regular work clothes are acceptable. Wool or heavy denim, such as blue jeans, works just fine for shielding radiation. Many professional weldors use leather aprons, jackets or pants for burn protection. And don't forget welding gloves. Your hands are closest to the heat and light source. Consequently, they are most vulnerable to burns from radiation, sparks and hot metal.

For maximum burn protection from your long-sleeve shirt, tuck the sleeves inside the top of the gloves and button the collar at the top. Many beginning weldors neglect to button up their shirt collar and end up with a nasty ultraviolet burn on their neck. Wearing a leather flap on the helmet to shield the neck and throat is the best way to avoid neck and throat burns.

Helmet—There are about 100 different safety-approved welding helmets, or hoods. I have two different helmets for myself.

Get the lightest helmet that covers your face completely. Whether you are welding bridges or ships for a living, or welding a lawnmower or child's bicycle, the lightest helmet will distract you the least. You need all the concentration you can muster to do a good welding job, so you certainly don't want a heavy helmet pulling down on your neck and head. Regardless of weight, though, you need maximum protection.

Arc-Welding Lenses—I prefer the rectangular 2 X 4-1/4-in. lens. It's available in various shades and tints. The standard tint is green #10. A darker tint for welding in bright sunlight is green #11 or #12. A #9 lens is OK for low-

hydrogen welding or TIG-welding steel. TIG or wire-feed welding aluminum requires a darker-tint lens, such as a #10, #11, or even #12. You should pick the darkest lens that still allows you to see the weld puddle.

Don't use gold- and silver-plated lenses. Although they are pretty, one tiny scratch in the gold or silver plating could admit enough ultraviolet light to burn an eye.

A tinted lens should always be protected from weld spatter, scratches and breakage. Cover it with a clear, disposable plastic or glass lens. Change the lens protector as often as necessary to assure distortion-free sight. Likewise, clean the lens as often as you would a pair of eyeglasses.

Gloves—Protect your hands and wrists from burns with leather gloves. Although leather doesn't burn easily, it will char and shrink if it contacts hot metal or a flame. Choose the most flexible gloves possible. For light TIG welding, I prefer deerskin gloves. For heavy arc welding on structural steel, heavy cowhide or horsehide gloves give you the protection needed.

Shoes—If you weld bridges professionally, you should wear leather, high-top, lace-up safety shoes. Regardless of whether you weld for a hobby or professionally, *never* wear low-cut shoes, particularly the slip-on type such as loafers. Sparks and molten metal can slip down inside your shoes. Regardless of your pain threshold, you'll drop *everything* so you can get your shoe off and the hot metal out while you dance the jig. And while you are dancing the jig, the dropped welding torch could start a fire or explosion! Never wear nylon jogging shoes. If exposed to high heat, they could melt to your foot! For hobby welding, an old pair of leather lace-up shoes will be OK.

Safety Glasses—If you can't afford safety glasses, you can't afford to weld. Don't think safety glasses are too much bother—they don't compare to the bother of impaired eyesight. Your eyes will not tolerate hot metal in them. My eyes much prefer to clearly see the beauty of my wife, so I wear my safety glasses when grinding, chipping, filing or anytime there's a chance of metal flying around.

OTHER PEOPLE'S SAFETY

When arc-welding, you have the obligation to protect bystanders from ultraviolet burns to their eyes. First of all, you should provide a suitable screen around the weld area so direct flash can't be seen by anybody not wearing correct eye protection. For a temporary screen, place a sheet of plywood or corrugated tin so it

RECOMMENDED WELDING LENSES

Application	Base Metal Thickness (in.)	Suggested Shade No.
Arc welding with		
1/16, 3/32, 1/8, 5/32-in. electrodes	1/8 to 1/4	10
3/16, 7/32, 1/4-in. electrodes	1/4 to 1	12
5/16, 3/8-in. electrodes	Over 1	14
TIG welding with non-ferrous 1/16, 3/32, 1/8, 5/32-in. electrodes*	Up to 1/4	11
TIG welding with ferrous 1/16, 3/32, 1/8, 5/32-in. electrodes*	Up to 1/4	12
Soldering	All	2
Brazing	Up to 1/4	3 or 4
Light cutting	Up to 1	3 or 4
Medium cutting	1—6	4 or 5
Heavy cutting	Over 6	5 or 6
Gas welding (light)	Up to 1/8	4 or 5
Gas welding (medium)	1/8—1/2	5 or 6
Gas welding (heavy)	Over 1/2	6 or 8
* For MIG welding, decrease shade no. by one		

Make sure your eyes are protected with the right lens when welding.

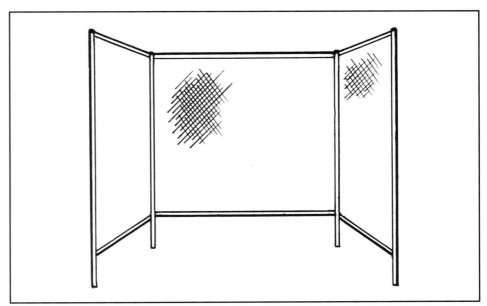

Be sure to electric-weld behind a welding screen to protect bystanders from ultraviolet radiation. You can make one from three pieces of 1/4-in. plywood, hinged together at the joints.

shields those nearby.

If you can see arc-weld flash from *any* distance, it can burn your eyes. The closer you are to the flash, the more severe the burn will be. The same rule applies to pets and other animals. Within reason, keep them from looking at the welding flash by using shielding. If anyone is in the weld-

I prefer one-piece goggles (left) for gas welding. They fit over eyeglasses, and don't get tangled like separate-lens goggles (right). Also, it's easier to get replacement lens for one-piece goggles.

Ruptured oxygen cylinder was overheated by a flame, causing it to explode. Photo courtesy of St. Phillips College, San Antonio, Texas.

ing area, *always* say "watch your eyes" before you strike the arc. This gives the person a chance to turn the other way or close their eyes. If you weld regularly in a certain area, permanent or portable screens should be used. These can be built or purchased from a welding-supply outlet.

GAS-WELDING SAFETY

Oxyacetylene welding presents its own set of safety considerations. Everything I said about shoes, gloves and clothing for arc welding also applies to gas welding.
Goggles—Goggles for gas welding or cutting are different than arc-welding helmets because they don't have to shield the weldor from hostile light. Oxyacetylene welding doesn't produce ultraviolet light, so it can't cause flash burns. The only burn you can get from oxyacetylene is from the heat of the flame, and from sparks from the oxidized metal being welded or cut.

The best goggles for oxyacetylene welding are fiberglass and have a 2 X 4-1/4-in. lens. You can wear most conventional vision-correcting eyeglasses under them. In addition, they are easier to put on and adjust than the separate-eyepiece goggles.
Bottled Welding Gases—Welding gases come in high-strength, high-

pressure cylinders. When full, they contain the following pressures at room temperature:

Oxygen—2200 psi
Argon—2200 psi
Hydrogen—2200 psi
Acetylene—375 psi

If a pressurized cylinder of oxygen or argon welding gas falls and breaks off a valve, the resulting release of high-pressure gas can turn it into a rocket. I know of one high-pressure cylinder with a broken-off valve that went completely through four automobiles and the brick wall of a welding shop. As with fire, bottled welding gas is useful when contained and controlled. But also like fire, pressurized gas can be destructive. *Always* chain welding-gas cylinders in an upright position to prevent them from falling. An acetylene cylinder must be kept upright to keep acetone, a stabilizing gas, at the bottom of the cylinder. If the heavier acetone gets into the cylinder regulator, it will damage its rubber parts.

Although acetylene is bottled at comparatively low pressure, the acetylene itself is highly explosive. Like leaking gas from a natural-gas pipeline, nothing happens if acetylene leaks into a confined space unless there is a spark or flame to ignite it. Then, it can explode just like a stick of dynamite. Never

carry flammable gas such as acetylene, hydrogen or MPS gas in the closed trunk of a car. If vapors escape, a major explosion could result. Instead, haul acetylene- or MPS-gas cylinders in an open truck or trailer.

Don't let these safety tips worry you unnecessarily. Every time you drive a car, you are carrying around enough gasoline to do about the same damage as the contents of any welding-gas cylinder. Therefore, if you observe proper safety precautions, you should have no trouble.

OTHER SAFETY TIPS

Work Area—The most common welding accident is burned hands and arms. Keep first-aid equipment for burns in the work area. And, as I already discussed, eye injuries can occur if you get careless. So keep handy a phone number for emergency medical treatment for unexpected injuries, particularly eye injuries. A medical doctor is the only one who can properly treat any eye injuries. Don't try to treat an eye injury yourself. Get to a doctor *immediately.*

The ideal welding shop should have bare concrete floors, and bare metal walls and ceilings to reduce the possibility of fire. Although you probably don't have

Cutting or grinding sparks will start fires just the same as sparks from a welder or cutting torch. Note operator wearing full face shield.

Inexpensive reading glasses keep you from having to wear your expensive prescription glasses while welding.

such a building, the important thing is to keep flammables, and rags, wood scraps, piles of paper and other combustibles out of the welding area. The same goes for wood floors; never weld or cut over one.

If you must weld in an enclosed garage, make every effort to eliminate anything that could trap a spark. Sparks can smolder for hours and then burst into flames. So, regardless of where you're welding, be sure to have a fire extinguisher nearby. Also, keep a 5-gallon bucket of water handy to cool off hot metal and quickly douse small fires.

Never use a cutting torch inside your workshop. Take whatever you're going to cut outside, away from flammables. Also, be aware that welding sparks can ignite gasoline fumes in a confined area.

Grinding Sparks—When grinding with a portable grinder or bench grinder, the resulting high-velocity sparks are tiny pieces of metal. These tiny projectiles are in an oxidized state and at the same temperature as metal being cut with a gas torch. Therefore, be as careful of where grinding sparks fall as you would if you were using a cutting torch.

Eyeglasses—An often-overlooked cause of less-than-perfect welding is not being able to see the weld

puddle clearly. Frequently, people who don't have good vision attempt to weld without eyeglasses. If you want to do a good welding job, but need glasses, have your eyes checked and corrected first. Or, wear your glasses.

If you don't want to subject your expensive bifocals to welding damage, there are other solutions. The least-expensive solution is to buy a pair of reading glasses at a drugstore or supermarket. These come in standard magnification increments of +1.25, +1.50 and so on. Welding supply shops also sell eyesight-correcting lenses that fit the 2 X 4-1/4-in. opening in helmets or goggles. These are available in +1.25, +1.50 and higher increments, too.

COMFORT

I suppose most people think a weldor at work is not comfortable. That is somewhat true, but the more comfortable you are while welding, the better your work will be. I prefer to sit when welding. Why stand and have leg fatigue interfere with concentration?

Just as you would sit down and get situated to write a letter, get yourself comfortable before starting to weld. Think about it. You wouldn't squat down and write a letter on the floor. Likewise, you wouldn't stand up with your arms

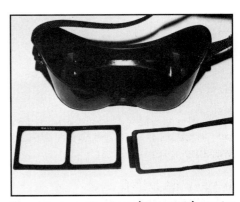

Special corrective lens (lower left) can be used in place of eyeglasses. Such lens are available at welding-supply stores. This lens fits either gas-welding goggles or an arc-welding helmet.

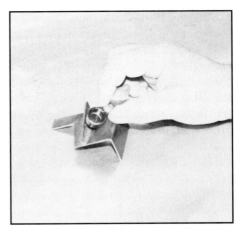

Using a 25X magnifying glass to inspect a fresh TIG weld. This glass, also called a *jeweler's loupe,* will reveal any cracks or porosity. Buy one at a hardware, welding-supply or jewelry store.

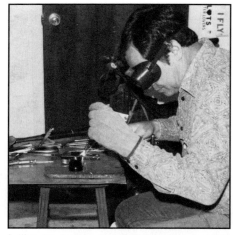

I like to be as comfortable as possible when gas welding. Two stools with a 3/8-in. steel plate make a small welding table. Old kitchen chair provides seating comfort. Photo by Lyle Sandell.

If you're going to do a lot of arc welding and oxyacetylene cutting, build a welding-and-cutting table like this. Plans for building it are on page 139.

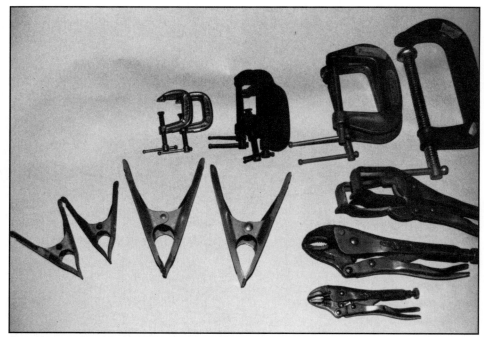

Assortment of weldor's helpers—various clamps for holding workpieces together—come in handy for all kinds of welding.

stretched out to write a letter with the paper against a wall. And you certainly wouldn't lay on your back, writing a letter using the side of the desk! If you tried writing in these awkward positions, your handwriting would not look good. That's why you can't weld as well lying on your back, or squatting on the floor.

There are exceptions. For instance, you can't flip a car on its roof just to weld on a new muffler. However, you can get yourself as comfortable as possible if you have to weld in such a position.

WELDING AREA
Chair—Get a lightweight, comfortable chair. I prefer a chair like that used by a draftsman—one that can be raised or lowered to suit the work height. The chair should have rollers so it can be moved without having to pick it up. Don't spend a lot of money. Check a used-furniture store or a flea market first.

Table—Build a metal-top welding table. The table top should be steel—not aluminum or wood—at least 1/4-in. thick and 2-ft square. Make the top larger if you plan many big, heavy welding jobs. A good all-purpose welding table uses a 3/8-in.-thick, 3 X 4-ft steel plate for the top. The frame to hold the welding table can be anything sturdy enough. For occasional welding, I simply lay my 2 X 3-ft steel table top across two 36-in.-high wooden shop stools. For a more permanent setup, consider welding a frame and legs made from angle steel or pipe to set the plate on.

Keep the table top fairly rust-free, especially if you are using it for arc welding or TIG welding. Rust is a poor electrical conductor and will even contaminate gas welds. I power-sand the top of my weld table occasionally to expose shiny steel. Never paint the top of a weld table because it blocks current flow from the ground elec-

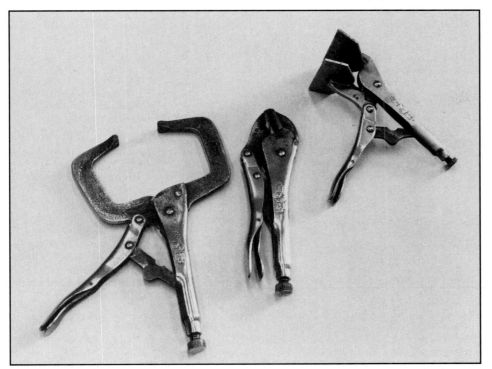

An assortment of Vise-Grip locking pliers should be in your toolbox. Each type has its special application. Photo by Tom Monroe.

Shield over C-clamp screw protects threads from arc-weld spatter. One glob of weld spatter can lock a C-clamp adjusting screw.

trode to the workpiece.

Drill a 1/2-in.-or-larger hole in the back side of the table top or frame and put a bolt and nut there for a more secure ground-connector attachment. See page 139 for a welding-table design you can build yourself.

Keep an assortment of C-clamps, weights, metal clothespin clamps and *holding fingers* nearby to hold the pieces in position while you weld. *Never* ask anyone to hold a small part while you weld it. They won't like it when the sparks fly and the metal heats up!

CONVENIENCE EQUIPMENT
Sandblaster—A great addition to any welding shop is a sandblaster. You can buy a small portable sandblaster that holds 50 pounds of sand for about the price of 3 tanks of gasoline. I used one to completely sandblast the frame of a 1956 Ford 1/2-ton truck I was restoring. The job took about 6 hours, but the result was a likenew frame that was ready to accept a coat of polyurethane paint.

A sandblaster is also handy for cleaning small parts and numerous projects. But it does take several

Sandblaster saves lots of chipping and wire brushing. Photo by Breven Finch.

minutes to get the sand ready. And after sandblasting, you'll need a bath to wash all the sand off. It gets everywhere—even where the sun doesn't shine!

If you buy a 4—8-cfm siphon-type sandblaster as illustrated on this page, you'll need at least a 1- or 2-HP air compressor. A 1/2-HP compressor doesn't have sufficient capacity. Pressure-type sandblasters work two or three times faster, but will cost up to ten times more. The siphon-type sandblaster will

Siphon-type sandblaster holds about 40 pounds of sand. It can also use other abrasives such as walnut shells.

Homemade screen is for sifting sandblaster sand for reuse. After third sandblasting, *silica 30* sand becomes too fine and will no longer clean parts adequately.

Pressure-type sandblasters do a better job than siphon types, but cost more.

Scrap metal comes in handy for practice, testing new weld joints and saving money. Identify each piece with a metal marker so you'll know what you have. Photo by Tom Monroe.

Ball-tip metal marker will mark just about any surface. This high-temperature marker, which can withstand 1100F (593C) before burning off, is great for indicating usable alloys of scrap pieces. Photo by Tom Monroe.

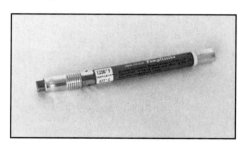

Temperature-indicating crayons are useful for monitoring maximum workpiece temperature. For example, I use 350F (177C) Tempilstik to indicate when desired preheat temperature is reached for welding aluminum castings. Photo by Tom Monroe.

Temperature-indicating paint works the same as temperature-indicating crayon—it melts when indicated temperature is reached. Temperature is shown on bottle. Photo by Tom Monroe.

do a good job of cleaning metal for the occasional, small welding job. For more frequent weldors, and more complex welded assemblies, consider investing in a pressure-type sandblaster.

Metal Marker—A handy item to have around the welding shop is a *metal marker.* It's a tube of paint with a ball-point end that marks almost anything. It will mark smooth metal, rough metal, oily metal, wood, glass or even plastic. You'll find these markers at most welding-supply shops.

I use a metal marker for identifying scraps of metal that are suitable for future use. A tube costs about as much as a hamburger, but lasts about two years in normal use.

Never use a scribe for marking metal. Scribe lines in metal act the same as perforations in a piece of paper—ready-made for tearing, breaking or cracking. If you do scribe-mark a piece of metal, the scribed area should be discarded because of the potential for cracks. Such a case would be marking a cut line.

Temperature Indicator—Often, when welding or heating metal, it's crucial to get the metal to just the right temperature—no more. Excess heat will often ruin the metal. This can happen even when loosening rusted or otherwise stuck parts, too.

The way to assure yourself of exact temperatures is with a temperature-indicating crayon or paint. Temperature indicator is applied to the area to be heated or welded. When the indicated temperature is reached, the crayon or paint melts like wax on a hot surface.

Temperature-indicating crayons and paints come in more than 100 different temperatures, from 100F (38C) to 2500F (1371C). Claimed accuracy is plus or minus 3F. These and other types of temperature indicators by Omega Engineering and Tempil can be purchased at most welding-supply stores. Or, write Omega Engineering, Inc., Box 4047, Springdale Station, Stamford, CT 06907 or Tempil Division, Big Three Industries, Inc., Hamilton Boulevard, South Plainfield, NJ 07080, for information.

3

What Kind of Welder is Best?

NASCAR Modified frame and roll cage, designed by Tom Monroe, is constructed of square, rectangular and round mild-steel tubing. TIG welding was used to join tubes. Photo by Tom Monroe.

Now that you know a little about the basics of fusion welding, soldering and brazing, and what special equipment and safety precautions are necessary, it's time to investigate the different types of welding outfits available and which ones are suitable for your needs. If you have decided to buy your own welding rig, start by making the right purchase. Don't spend money on a piece of equipment you cannot use.

In this chapter, I discuss seven types of welders I think are appropriate for the small welding shop or do-it-yourselfer:

Oxygen/propane
Oxygen/acetylene
(oxyacetylene)
A-c current arc
D-c current arc
Heliarc, or TIG
Wire-feed, or MIG
Spot

There are several other types of welders. I discuss plasma-arc welding and cutting in Chapter 10. A plasma-arc welder is similar in design and usage to a TIG welder. The advantage of plasma over TIG is its highly constricted arc, which reduces the heat-affected zone in the base metal. Plasma-arc cutting, however, is quickly gaining acceptance by auto-salvage-yard operators for dismantling cars and trucks. It is also used in body shops because it quickly cuts through painted and grease-covered metal.

The seven types of welders shown in the photos vary widely in cost. The oxygen-propane torch is

Joe Ruttman, in sister car to that shown above, navigates the infield road course at Daytona International Speedway.

19

My friend Rusty is fusion-welding sheet metal with an oxygen-propane torch. Oxygen-propane setup is least expensive of all metal-fusing rigs, but has limited use because it produces a very small amount of heat for a very short time.

I constructed this Fokker Tri-plane replica fuselage using 4130 steel and oxyacetylene welding. Fuselage weighs less than 100 pounds. You should be able to do the same providing you master fitting and jigging, and oxyacetylene welding.

the least expensive. Most expensive are the TIG, plasma-arc and wire-feed welders. A spot welder suitable for welding aluminum can cost $50,000.

The descriptions of the various welders should help you decide which welder is best suited to your needs and financial situation. Once you decide which welder you need, read the appropriate chapters to learn more about its setup and use.

OXYGEN-PROPANE TORCH

Notice this is called a *torch,* whereas the other rigs are called *welders.* The oxygen-propane torch is an inexpensive unit that will produce a maximum flame temperature of 5300F (2927C), but only for a short time. After about 15 minutes of maximum output, the oxygen cylinder is depleted.

Two cylinders are used, one for oxygen and one for propane. Each cylinder costs about $10—about the same price to refill an 80-cubic foot (cu ft) *Q-size* oxygen cylinder for an oxyacetylene welder, page 43. The oxygen should last for more than a full year of home-project work. The initial cost of an 80-cu ft oxygen cylinder is high—more than $100—but this is a once-in-a-lifetime expense because the cylinder is refillable. By

the way, this 80-cu ft cylinder must be replaced at the end of 100 years as a safety measure.

You will also find, if you buy a small oxygen-propane torch, that there is a definite limit to what it will weld in its 15-minute duration. That's just about enough time to weld a cracked bicycle frame—no more. If you want to weld something such as a 1 X 1/4-in.-wide steel brace to your boat trailer, an oxygen-propane torch just won't do it. There are only so many thermal units of heat available.

About all you'll be able to do with an oxygen-propane torch are jobs like brazing or welding cracked bicycle frames, soldering copper tubing for home-plumbing projects and silver-soldering handles back on stainless-steel cookware. You can also do some light heating for metal forming, such as heating 1/4-in.-diameter steel rod for bending and similar projects. If you are doing low-temperature jobs such as soldering copper water pipes in a house, the gas in the cylinder will last about eight hours.

OXYGEN-ACETYLENE WELDER

Generally speaking, the oxygen-acetylene gas welder is the most versatile of all welders. If you can

Oxyacetylene is the most useful general-purpose welder. Regardless of the welding shop, oxyacetylene should be included.

only afford one welder, buy an oxyacetylene rig. The parts necessary to begin welding cost about $500, but you can lower this cost by 75% by leasing or renting the oxygen and acetylene cylinders for a few months. If you buy the welder and lease the cylinders, the oxygen-acetylene welder is one of the least-expensive welders to use.

With an oxyacetylene welder, you can do many projects in a home workshop or even a commercial welding shop. Years ago, before TIG welding was invented, tubular-steel airplane fuselages were completely welded with oxyacetylene.

The airplane-fuselage frame shown on page 20 was fusion-welded with an oxygen-acetylene welder. An oxyacetylene welder is capable of welding steel and aluminum over a wide range of thicknesses—from paper-thin to about 1/4 in. Thicker metals should be welded with an electric-arc welder.

Welding Steel—A gas welder can be used to weld almost any steel part—up to 1/4-in. plate or 1-in.-OD bar stock. It can also be used to cut out pieces of metal. And you can use a gas welder to tack-weld metal pieces to hold them in place for arc welding.

Cutting Steel—One of the fastest and smoothest methods to cut steel bars, plate and angle into usable shapes is with an oxyacetylene cutting torch. Almost all gas-welding kits include a cutting-torch attachment. The average cutting torch will cut through steel from paper-thin up to 1-in. thick. A regular-size cutting torch can also cut through a 2-in.-thick axle shaft. Arc welders are capable of cutting through steel sheets up to 1/4-in. thick, but the accuracy of a *carbon-arc* cutter leaves a lot to be desired. The gas-cutting torch is the best.

Brazing—Brazing is a metal-joining operation that is done almost exclusively with a gas welder. The brazing process, explained in Chapter 7, is used where a strong joint is desired without the high heat associated with fusion welding. Most bicycle and tricycle frames are brazed. Even some race-car frames are brazed.

Silver & Lead Soldering—The gas welder is also a perfect tool for silver and lead soldering. Whereas lead soldering is satisfactory for low-temperature applications such as water pipes, silver soldering is useful in high-load, high-

Completed Fokker Tri-plane replica incorporating fuselage shown on previous page.

I brazed this 0.049-wall, 1-in.-OD mild-steel race-car frame. Car won SCCA Divisional Points Championship in 1969, 1970 and 1971.

Completed race-car with frame shown above. Power is from rear-mounted, two-stroke, three-cylinder SAAB engine and four-speed transmission. Richard Finch topped 140 mph racing at Phoenix International Raceway. Photo by Jim Pittman.

temperature applications such as joining aircraft parts and stainless-steel kitchen cookware.

Other Uses—The basic function of a gas welder is heating. Oxyacetylene can be used for many types of welding, but it has other uses, too. Its smaller tips will heat metals for heat-treating, for melting lead and even for heating nails red hot to pierce holes in wood.

A much larger, special *rosebud* tip can be used where a lot of concentrated heat is needed. Such an

Oxyacetylene cutting torch such as this can cut steel up to about 1-in. thick. I use piece of angle steel as a guide to cut straight line. Photo by Breven Finch.

Rosebud tip (arrow) is used for heating large areas for forming, bending and preheating.

Using rosebud tip to free stuck U-joint flange from axle stub shaft. While puller applies force, heat expands flange to break-loose axle shaft.

Most common of small-welding-shop or home-workshop rigs, buzz-box arc welder is so named because it makes a buzzing sound when operating.

example is preheating large pieces of metal prior to welding and for heating metal for bending and forming. In Chapter 15, there are several projects that require heating and forming metals.

Any oxyacetylene-welding tip can be used for heating rusted, stuck or *frozen* parts to aid disassembly. Just pick the proper-size tip. It should be large enough to heat the parts cherry red, but small enough to avoid melting them.

Similarly, an oxyacetylene welder can be used to assemble parts, such as press-fit wrist pins in an engine. The wrist-pin end of a connecting rod is expanded by heating it. Before it cools, the wrist pin is slipped through the piston and rod. As the pin and rod approach the same temperature, the rod contracts and traps the pin with an *inteference fit.*

ARC WELDER

Shielded metal-arc welding (SMAW), as it's called in the industry, or *arc welding,* as it's commonly known, is the most common, least expensive and most flexible electric-welding process. There are two basic types of arc welders: a-c current and d-c current. There are also combination a-c/d-c current arc welders.

A-C Arc Welder—An a-c-current, or *buzz-box,* welder is easy to set up and use. Although some light-duty 110-volt arc welders are available, I recommend a 220-volt unit. It only has to be plugged into a 220-volt, a-c-power source and turned on. Most commercial shops are wired for 220 volts, as is any residence using a 220-volt electric stove or clothes dryer. However, few places actually have the proper 220-volt receptacle to plug one into—see the illustration, page 74. Often, a 220-volt receptacle either is not located in a convenient place or is already hooked up to a stove or dryer. The cost of providing an additional outlet for a buzz-box welder often equals or exceeds the cost of the welder.

If your shop or home is wired for a 220-volt stove or dryer, you can make an adapter to plug a 220-volt buzz-box into the existing outlet. Buy a welder receptacle, a stove/dryer 3-prong plug and make up an extension cord using 8/3S (8-gage, 3-conductor, flexible) cord. Obviously, you have to perform your arc-welding projects when the stove or clothes dryer isn't in use.

As its name implies, a-c current *alternates* current polarity—direction of flow—from positive to negative at a specific frequency. In the U.S., it's 60 cycles per second; in other countries, 50 or 90 cycles per second.

An a-c buzz-box welder is best suited to welding 1/8—1/2-in.-thick mild steel. It will weld thicker metals, but if many inches of weld are to be deposited, a heavier-duty a-c/d-c welder would be better. A buzz box will handle most home arc-welding projects as well as being ideal as a commercial auxiliary welder.

An a-c arc welder will do a good job of welding a utility-trailer frame, automobile frame, and so on. It's also good for building shop equipment and doing many kinds of welding repairs on thicker metals.

In order to be effective, a buzz-box should always be backed up by an oxyacetylene welder and cutting torch. The torch is used to cut pieces of metal to the correct size and shape as needed. An expensive alternative to the cutting torch is a power hacksaw, metal shear or similar piece of metal fabricating equipment, as detailed in HP Books' *Metal Fabricator's Handbook.*

Variable or Fixed Amperage?—Is a variable-amperage or fixed-amperage buzz-box welder best? With a variable-amperage welder, you can adjust it a little hotter or colder to suit each job. But do you really need the variable-amperage capability?

As you gain experience with your buzz-box, you'll learn that 75 or 90 amps welds almost everything. A typical fixed-amperage buzz-box has these settings: 40, 60, 75, 90, 100, 115, 130, 145, 160, 180, 200 and 225 amps. At the 1/8-in. rod-size settings, you are within 5—7-1/2 amps of the *perfect* setting. And you can repeat the setting every time. With a variable-amperage welder, you never get the same setting twice in a row—even if you leave the infinitely variable control at the exact same setting. So, don't spend the extra money for a

D-c welders are usually found in welding shops where thick, heavy parts are welded. Such welders will operate for long periods of time without losing power or overheating.

Portable d-c generator welding rig mounts on truck or trailer for welding in field. Photo courtesy Lincoln Electric.

variable-amperage welder.

Rod Considerations—Special arc-welding rod, or *stick electrode,* must be used with the a-c arc welder. This rod will not store indefinitely. Its coating absorbs moisture from the air, deteriorating the shielding needed for a good weld. Special storage precautions must be observed to protect arc-welding rod, pages 74—75 and 121.

D-C Arc Welder—A d-c arc welder is usually considered the heavy-duty machine of welding. It's commonly used in production welding shops and schools where the welders are expected to operate 8—12 hours a day without stopping. A d-c arc welder usually requires a special 220-, three phase or *440-volt* a-c outlet for operation. Therefore, it cannot be wired in as easy as an a-c welder.

The d-c, or *direct-current,* feature provides the ability to switch from *positive* to *negative polarity* as required for welding different metals. This results in better control of the weld heat.

There are d-c arc welders other than the wired-in type. Usually, a gasoline or diesel engine drives a generator. The generator produces direct current that may be used *as is* for welding power. The wired-in machines may operate off a-c cur-

rent that is passed through a *rectifier circuit* or one-way circuit, to convert it to d-c welding current.

Most welders have the provision to switch polarity, or to switch to a-c current, which has no polarity provision.

A d-c-current arc welder, best suited to large commercial shops or schools, costs four to 10 times as much as a buzz-box welder. There are small a-c/d-c arc welders designed for small jobs. However, they cost about twice that of a buzz-box welder.

A d-c welder allows the operator more precise control than the buzz-box. It also uses stick electrodes giving a relatively coarse weld. D-c welders are generally used for welding pipe, heavy farm machinery and other such jobs. With certain adapters, they can be used for TIG welding, which produces a smooth, precise weld.

HELIARC (TIG) WELDER

A TIG (Tungsten Inert Gas) welder, commonly known as a *Heliarc* welder, is the most precise of all hand-operated welding machines. It requires considerable setup work. It must also be maintained regularly or prepared for storage after use because of water-cooling tanks, argon tanks and their hoses. TIG welders are used

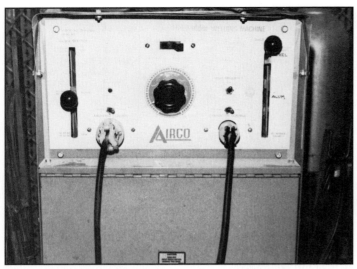

TIG welder produces pretty welds in steel, aluminum, stainless steel and titanium. It is normally used in aircraft, missile and race-car welding.

Left-handed student weldor is learning to use fully-equipped TIG welder. Torch is usually held with writing hand for maximum dexterity. Student will get ultraviolet-ray burns on arms because they are not covered with long sleeves.

in the nuclear-power industry for precise repair work where defects in conventional welds have been found. They are also commonly used in the aircraft and space industries.

As with the large d-c welder, a TIG welder is usually wired in permanently in a welding shop. It can be made portable by plugging into 220 volts, 440 volts or by providing a gasoline- or diesel-powered portable generator. TIG welders can be set up several different ways as detailed in Chapter 9. Here are three of the more-common setups:

Setup 1: A-c/d-c with high-frequency arc stabilization, water-cooled torch, foot-pedal amp control and flow-controlled argon shielding gas. With this versatile setup, you can weld with stick electrodes, allowing the use of the foot-operated amp control for some interesting arc-welding effects. You can start with a *hot* arc for easier starting and taper off to a *cold* arc for thin spots or more accurate filler-metal deposit.

Setup 2: D-c welder with TIG torch, argon gas and provision for stick-electrode welding. This set-up can be used to TIG-weld steel, but its capabilities are restricted to steel with a narrow range of metal

thicknesses.

Setup 3: Same as setup 2 except added high-frequency arc stabilization. In this case, the high frequency provides for a cleaner, smoother weld.

Parts welded with a TIG-welder are usually very precise assemblies such as race-car, airplane or missile parts. Much of the nuclear power-plant industry uses TIG welding. A complete TIG-welding machine can cost $2500 — $4000.

WIRE-FEED (MIG) WELDER

A MIG (Metal Inert Gas) welder takes about the same amount of setup effort as a TIG outfit. Although the two contain similar equipment, a MIG welder does not use water cooling. In place of it are wire-roll holders, automatic wire feeders and a wire gun that is used somewhat like a stick-welder electrode holder. MIG-weld quality is between that of stick welding and TIG welding. Its major advantage is that the operator can weld continuously, without stopping until the roll of wire filler material runs out. Because most wire rolls last for eight or more hours of continuous operation, the operator is the only limiting factor. He just has to stop and rest every few minutes. In

comparison, the arc weldor must stop for a new welding rod every 30 seconds; a TIG weldor has to stop about every 60 seconds to get a new piece of filler rod.

Clearly, the MIG, or wire-feed, welder is a high-production machine. It has been adapted to robots for automobile manufacturing as well as many other high-production jobs. Robots do not have to stop and rest, so the welds can be done more quickly and continuously.

Wire-feed welders are not used where precise welding is required. Why? Because the degree of operator control is not much greater than that of a stick-electrode welder. I once designed a wire-feed-welding setup to weld airplane seat frames. Although the seats and welding methods were certified and approved by the Federal Aviation Administration (FAA), the wire-feed welder was not precise enough to weld airplane engine mounts!

Wire-feed welders are capable of welding aluminum structures, horse trailers, bulldozer frames, pipe joints and similar high-volume work. They have no place in the average home workshop, but are ideal for race-car fabricating, and auto-body shops.

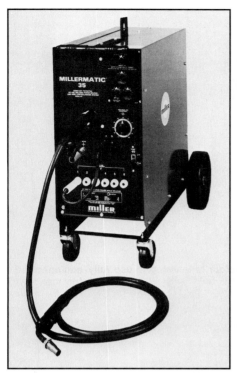

MIG welder is champion of high-production shop welders. It is easiest welder to use, and produces nice-looking, high-quality welds. Photo courtesy Miller Electric Mfg. Co.

Robot MIG welder is capable of continuous welding without stopping for welding rod—or coffee breaks! Photo courtesy of Hobart Brothers Corporation.

SPOT WELDER

A spot welder is limited to sheet-metal welding. It is ordinarily used in production welding shops for joining large quantities of sheet-metal parts. The auto industry has used spot welding for many years to assemble car bodies.

More recently, portable spot welders that connect to the electrode holder of an arc welder with 50-amp minimum output have become available.

WHERE TO BUY WELDING EQUIPMENT

You can do a lot of armchair research, perusing the welding-equipment mail-order catalogs. Some mail-order outfits sell name brands of welding equipment as well as their own. Look at the specifications and take the time to think about the equipment you really need.

You can also try retail outlets such as Montgomery Ward and Sears, especially if you want to buy on credit. This also gives you

a chance to inspect the various features of each machine.

Before buying, visit a vocational technical school to see what welding machines they use. Ask the instructors for their opinions about the equipment. They should be able to tell you which welders offer good service and which ones do not.

Look in the yellow pages of your telephone directory for welding-supply stores. Visit several of these and find out about "service after the sale." If you can't buy spare parts for it, don't buy that welder.

Never buy a welder without trying it out, particularly an electric welder. At least have it demonstrated for you. For instance, if you plan to build a trailer out of angle steel, take a some pieces with you and ask the salesman to make some welds while you watch—with a welding helmet on, of course. If the salesman tries to sell you one of the little 110-volt

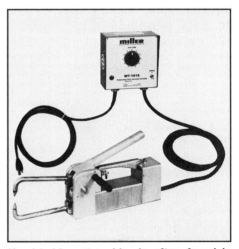

Hand-held spot welder is often found in sheet-metal shops. It is for welding thin sheet steel lapped two thicknesses, such as found in car bodies. Various tip and mounting configurations are available. Photo courtesy Miller Electric Mfg. Co.

arc-welding machines, you'll discover right away that you need something more substantial. Watch out for gimmicks like arc-welders that will "work" off a car battery, and 110-volt welders that are incapable of welding a tricycle frame.

25

Fitting & Jigging

4

Fixture built from 10-in. wide-flange beams supports late-model stock car being built by Kevin Rotty in his home workshop. Supporting screws are used for leveling fixture. Note material racks on wall, TIG welder and good lighting. Tubing notcher is mounted on fixture in foreground. Photo by Tom Monroe.

Fitting is the process of cutting and shaping metal pieces so they fit together without large gaps. Big welding operations usually employ a weldor-and-fitter team—two people who fit and complete the welds. If you were to time and apportion the operation, fitting takes about 80% of the time; welding takes the balance. These percentages illustrate how important fitting is to making good welds.

You must fit pieces properly *before* welding to make sure they stay in place during welding and subsequent cooling. The idea is to avoid large gaps that must be filled with weld bead.

Jigging is assembling a project in a fixture to ensure that the welded assembly conforms to design specifications. Jigs are used in large-scale production to assure consistent quality. You must plan ahead of doing the actual welding in order to get good welds.

FITTING

Thin-Wall Tubing—When welding thin-wall steel tubing, such as for a race-car roll cage or an airplane fuselage, I fit the tubing joints so there are no gaps—the joints are almost watertight *before welding!* At the very least, your welding project's joints should fit close enough so a welding rod cannot be inserted between them. See the photo, page 27.

One of the toughest fitting jobs is building a set of tubular engine-exhaust headers. For welding, fit the weld joints as closely as possible. It's not easy. Even though I've had a lot of practice, I sometimes overtrim a curved section of exhaust. If I overtrim a tube, I have to splice in a section or start over with a new piece. Practice is still the best way to get good fits.

Always cut the pieces a little longer than necessary so you'll have plenty of metal to file and fit into the proper shape. As one of my welding supervisors used to say, ". . . you cut it off twice and the piece is still too short!" This means that if you start with a metal piece that's too small, you'll never be able to fit it properly by trimming. So, be sure to *rough-trim* a part first to provide extra metal for final fitting. The resulting fit will be much more accurate.

Fishmouth Joints—I've tried

Fixture for spot-welding Bricklin sheet-metal body structure pivots so all weld seams are easily reached by simply rotating the fixture. Chances are you'll never need a welding fixture this sophisticated. Photo by Tom Monroe.

Simple plywood-and-nail jig for 4130-steel-tube airplane frame. Fit tubing so 1/16-in. welding rod cannot be inserted in weld-seam gap.

Tubing notcher, such as this one from Williams Low Buck Tools, speeds fitting one round tube to another. Photo by Tom Monroe.

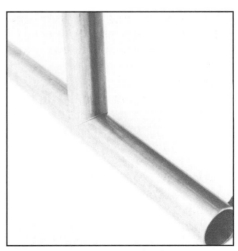

Correctly fishmouthed tube has no gaps. This will help ensure the best possible weld. Photo by Tom Monroe.

numerous methods to make *fish-mouth joints*—the half-round ones used to butt one round tube into another—faster and easier. Still, the easiest way for me is to mark the tube in the approximate angle of the desired fit, then cut it slightly oversize with a saw. This allows the fishmouthed tube to extend slightly past the centerline of the tube to which it is being fitted.

After sawing the tube slightly longer, I file and fit, and file and fit until the fit is just right. You can use a thin grinding wheel to do the rough fitting. Only hand-filing assures a perfect fit. When the fit is satisfactory, file off all burrs so they won't overheat and contaminate the weld. Also, removing the burrs makes a better-looking weld.

When I was welding for Aerostar Airplanes, I built more than 100 engine mounts with fishmouth joints. All passed Magnaflux inspection. One of the main reasons for this was tight-fitting weld joints. A little extra preparation pays off in better welds.

Angle & Plate Steel—Even when fitting angle and plate steel, you need a tight fit. Any wide gaps will vary weld quality, making the finished joint weaker.

Heavy-Gage Material—When fitting thick-wall tubing, heavy pipe or steel plate, you must bevel the edges to provide proper penetration, page 30. Some certified-welding specifications call for the gap to be less than one-third the diameter of the welding rod being used, regardless of base-metal thickness. That's a tight fit!

Tools—Tools for fitting range from the ubiquitous hacksaw to a heavy-duty power shear. The hacksaw is obviously a handtool; the shear a power tool. You'll need some tools from each category to do a quick and accurate job of fitting. You'll also need some tools to complement your cutting tools.

Measuring tools are needed to establish where a cut will be made. One of the handiest measuring tools is a retractable steel tape. A

Carpenter's square is useful for laying out parts and setting them up for welding. Photo by Tom Monroe.

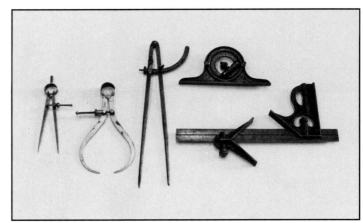

Dividers, calipers, machinist's square, protractor and centering tool are necessary for doing accurate layout and fitting work. Photo by Tom Monroe.

Power metal shear is a great labor and time saver. If you have a lot of metal to cut, try to do it in an evening class at your local college or high school. You could also have it cut at a steel-supply store for a fee. Photo courtesy of St. Phillips College, San Antonio, Texas.

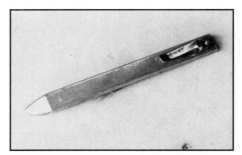

Soapstone in holder goes in shirt pocket for quick access. Bench grinder is great for sharpening soapstone, if you can tolerate white dust. Photo by Tom Monroe.

10-ft tape will usually do; however, big projects can require a 50-ft tape. For smaller jobs, you can use a 12- or 18-in. rigid steel rule. A 6-in. flexible rule in your shirt pocket can come in handy.

Finally, a carpenter's framing square is necessary for laying out those square cuts. One of these durable tools is great for marking parts to be cut or fitted, or setting them up to be welded.

In addition to a framing square, you should also get a machinist's combination square. Use this tool for fine layout work and for work involving angles other than 90°.

Marking tools are needed to provide a line for making cuts ac-cording to your measurements. When making a cut with a torch, you'll need a mark that won't be obliterated by the flame. These marking tools include a scribe and a center punch or *prick punch*. A center punch will work, but a prick punch is made specifically for this purpose—making a line of closely spaced prick marks.

Another method for marking metal to be cut with a cutting torch is with a soapstone marker. It marks like chalk, but will not burn at high cutting temperatures. By all means, have several pieces of soapstone marker in your toolbox.

For marking arcs or circles, you'll need a *divider*—a special scribe that's similar to a compass, but with two sharpened steel points. One place you might want to use a scribe mark and blueing is for flame-cutting flanges for ex-haust manifolds. Here, the cut must be precice, and the scribe line is better than soapstone for accuracy.

Remember that you should only use a scribe for marking a cut line. This is particularly true when marking sheet metal because of the tendency of a tear or break to start at a scribe line—the scribe line creates a stress riser.

Handtools include many tools that inhabit the typical toolbox. As mentioned previously, the hack-saw is the first to consider. Just make sure you have plenty of re-placement blades. A hacksaw is great for cutting tubing and small structural shapes such as angle iron and channels. Tinner's snips are great for making straight cuts on sheet metal; make curved cuts with aviation snips. Aviation snips can also be used for trimming thin-wall tubing.

For final fitting and smoothing, you'll need an assortment of files. Start with three basic styles: flat, half-round and round. Get double-cut, coarse-tooth files for remov-ing the most material with each pass. As for the size of the files, bigger projects require bigger files, and vice versa.

Power tools remove material quickly. Professional weldors pre-fer power tools because time is money. Even though the initial tool cost is higher, time is worth more. If you weld much angle steel, a 6-in. rotary, hand-held grinder is invaluable. Your welds can look totally professional if you dress them with a grinder. A grinder is also great for fitting and dressing after using the cutting torch.

There are many power tools from which to choose. One of the most common is the disc grinder. It's great for smoothing cuts, especially those made by a torch. A disc sander can be fitted with a cup-type stone, which removes metal quickly. For smaller fitting and smoothing jobs, a 9-in. grinder works particularly well. It's easier to control; thus, a more accurate fitting job can be done.

Die grinders—pneumatic or electric—are useful for making accurate fits. Abrasive stones or carbide cutters can be used, depending on the material being fitted. For instance, a carbide-steel cutter is useful when fitting soft metals, such as aluminum, because of the tendency of an abrasive to *load up*—clog with metal particles. A cutter will also do this, but it is much easier to clean. When coated with *paraffin wax,* a cutter has less tendency to load up.

For making fast cuts, a saber saw can be used in place of the hacksaw. The powered, oscillating saw rapidly eats through thin-gage metal, and must be used with care. It takes more effort to follow the cut—line than to make the actual cut.

For making cuts on heavy-gage steel, it's frequently best to use a cutting torch, page 59. Because flame-cutting leaves a relatively rough edge—how rough depends on the skill of the operator—it's always necessary to do the final fitting and smoothing with a grinder.

Safety—When using cutting and trimming tools, protect yourself against flying metal chips. This is especially important when working with power tools, such as high-speed grinders and cutters. Protect your eyes. Clear goggles are OK, but a full face shield is better.

Don't forget your arms and legs. Wear a shirt and pants with full-length sleeves and legs. Wear gloves when using a disc grinder or torch.

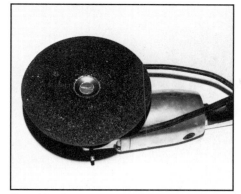
Portable electric grinder is useful in an arc-welding shop. It's great for cleaning up edges that have been cut with an oxyacetylene torch.

This beats a hacksaw. Portable electric band saw is used to trim crossbar of wrought-iron fence. Band saw is plugged into portable welder, which supplies a-c power. Photo by Tom Monroe.

WELDING JIGS

A *welding jig* is a fixture designed for holding parts in position during welding. The term scares many would-be weldors because they envision a very complicated device designed by a welding engineer. This is true only if it was designed to weld hundreds of parts on a production-line basis. *Jig-welded* usually means that all *weldments*, or welded assemblies, come out looking exactly the same, with consistent quality. For example, you should use a welding jig if you're building 1000 airliner seats.

For a one-time project, a jig can be as simple as Vise-Grip pliers or just a piece of angle iron used to prop up a part while it's welded in place. The more sophisticated welding jig can be like that shown below for welding race-car frames and roll cages. Fortunately, you

Welding jig on 20,000-lb surface plate supports stock-car frame and roll-cage parts while they are tack welded. Assembly is then removed and finish-welded on floor. Photo by Tom Monroe.

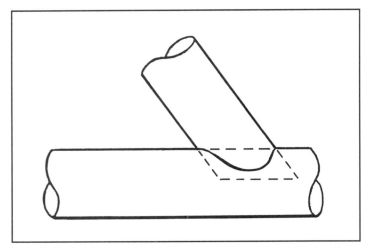

As indicated by dotted line, I fishmouth steel tubing by first making it slightly longer than necessary. I then cut and grind or file it to fit tightly.

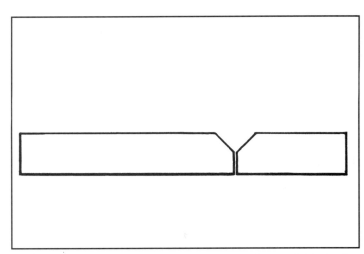

Bevel weld seam for better weld penetration if material is thick. Make several weld passes to fill bevel.

probably won't have to concern yourself with such a device. The most frequently used jig in my welding shop is a simple *three-legged finger* like that shown on page 137. The metal finger takes the place of your finger so you won't get burned while holding a part in place during welding.

Remember, a jig is whatever it takes to hold parts in place until *tack-welding* and finish-welding can be done. Tack welds are a series of very short welds spaced at even intervals. The tack welds hold two pieces of metal together so they can be finish-welded.

Wooden Jigs—A wooden jig is just that—plywood or particle board with small wood blocks nailed to it to hold metal pieces in place. When welding one or two assemblies, such a simple wooden jig is sufficient. See the illustrations, page 31. I have welded airplane parts, race-car parts, even factory production parts such as turbocharger wastegates, turbocharger-control shafts and seat headrests on wooden jigs.

Permanent Steel Jigs—Factories use heavy steel welding jigs to assure consistent sizes and fit of parts. You wouldn't want to buy a new exhaust system for your car and discover that the factory had welded the muffler-inlet pipe on the wrong side. Factory welding jigs assure that welded parts will be interchangeable. Even small

Weldors take a break in automotive frame-welding shop. Gimbaled jigs with clamps are used to retain frame pieces while they are MIG welded. Photo by Tom Monroe.

race-car factories have welding jigs to assure interchangeability and to improve production rates. See photo on page 32 of a weldor at Le Grand Race Cars welding a formula-car A-arm in a steel welding jig.

Welding jigs have their drawbacks, though. For example, when welding 4130 steel in a heavy welding jig, the parts sometimes must be *stress-relieved* to remove internal loads. The jig doesn't allow the parts to twist and conform to stresses from warpage. It holds the weldments in position, regardless of how they want to move. Therefore, internal stresses develop in the welded assembly. These stresses must be relieved, or the part may fail when it's put into use, or *service*.

STRESS-RELIEVING WELDED ASSEMBLIES

When a complicated, rigidly braced structure such as an airplane engine mount, power-plant high-pressure steam pipe or a race-car-suspension member is welded, stresses remaining in the metal can cause premature fatigue

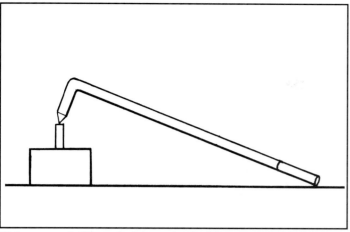

Make several different-size mechanical fingers to hold pieces in place while you tack-weld them together.

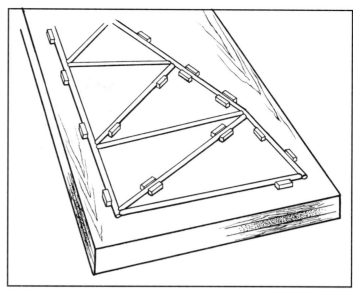

Welding jig for fuselage of steel-tube airplane: I use 1-in.-thick particle board and nail 1 X 1-1/2 X 2-in. pine blocks to it to hold tubing in place. Position blocks about 3 in. from each weld joint to avoid fire hazard.

cracking—caused by many loading and unloading cycles—unless they are relieved. Stress-relieving is accomplished by heating part or all of the structure to about two-thirds of the melting point and then cooling it slowly. This allows the molecules in the structure to relax and stay relaxed. See the chart, page 7, for melting temperatures of common metals.

However, stress-relieving must be done correctly. The fact is, more damage can be done by incorrect stress-relieving than by none at all.

Ideally, 4130 is stress-relieved by placing the welded assembly, still in a weld jig so it will hold its shape, into a large oven and heating it to 1050F (566C). At that temperature, both the *steel* and the big *weld jig* will glow blood red! I think you can see right away why stress-relieving is not easy to do!

Another way to stress-relieve 4130, and certainly my last choice, is to heat the area of the weld joint to 1250F (677C) with an oxyacetylene torch *in still air*. After playing the torch—moving it back and forth—over the weld joint for several minutes, back it away slowly, over a period of 2—3 minutes.

When stress-relieving, be ex-

tremely careful that no cool air hits the bead. A cool-air draft over a hot weld would be like throwing cold water on a hot light bulb. It will break. Another big problem is trying to simultaneously apply even heat to all sides of the weld joint—nearly impossible with just one torch! Because this stress-relieving procedure is so hard to do, I don't try it myself.

The decision to stress-relieve or not should be determined by the intended use of the welded part or assembly. For example, I did not stress-relieve some 1600 TIG-welded mounts for 300-HP airplane engines, and there was not one case of structural cracking! I'm not saying that stress-relieving is no good, just that it isn't always necessary. The aforementioned engine mounts shouldn't experience shock loading that is common and critical with some welded assemblies such as a sprint-car roll cage.

Gas-Welded Structures—When torch-welding an airplane fuselage or race-car frame, I stress-relieve the just-completed weld by slowly pulling the torch away from the work. I do this over a period of about 60 seconds. *I never just finish a weld and jerk the torch away.* That would surely cause cracking.

Another way to minimize stress cracks in welded assemblies is controlling the air temperature in your workshop. Never weld in a cold or drafty workshop. *A weld is more sensitive to cold and drastic temperature changes than the human body!* You'll have the best results welding in a room temperature of 90F (32C), but 70—80F (21—27C) works OK, too. All you'll have to do is become accustomed to working at above-average room temperature.

Don't try to weld in extremely cold weather. The chances of 4130 steel cracking after welding at 40F (4C) are 20 times greater than at 80F (27C). Mild steel is less prone to cracking from cold-air shock after welding.

Temperature-Indicating Crayons or Paint—If you're working on a project where stress-relieving is necessary, you can arrive at the precise temperature for stress-relief by using temperature-indicating crayons or paint such as those from Tempil or Omega, page 18.

As previously discussed, temperature indicators are applied to the critical area that's to be heated. When the indicated temperature of the crayon or paint is reached—for example, 1250F

TEMPIL TEMPERATURE INDICATORS

F	C	F	C	F	C
100	38	306	152	977	525
103	39	313	156	1000	538
106	41	319	159	1022	550
109	43	325	163	1050	566
113	45	331	166	1100	593
119	48	338	170	1150	621
125	52	344	173	1200	649
131	55	350	177	1250	677
138	59	363	184	1300	704
144	62	375	191	1350	732
150	66	388	198	1400	760
156	69	400	204	1425	774
163	73	413	212	1450	788
169	76	425	218	1480	804
175	79	438	226	1500	816
182	83	450	232	1550	843
188	87	463	239	1600	871
194	90	475	246	1650	899
200	93	488	253	1700	927
206	97	500	260	1750	954
213	101	525	274	1800	982
219	104	550	288	1850	1010
225	107	575	302	1900	1038
231	111	600	316	1950	1066
238	114	625	329	2000	1093
244	118	650	343	2050	1121
250	121	675	357	2100	1149
256	124	700	371	2150	1177
263	128	725	385	2200	1204
269	132	750	399	2250	1232
275	135	800	427	2300	1260
282	139	850	454	2350	1288
288	142	900	482	2400	1316
294	146	932	500	2450	1343
300	149	950	510	2500	1371

Use temperature-indicator chart for converting Fahrenheit to Centigrade or vice-versa. Chart indicates availability of Tempilstik and Tempilaq temperature indicators. Chart courtesy Tempil Division, Big Three Industries, Inc.

After carefully fitting pieces in steel jig, weldor brazes front-suspension A-arm in about 20 minutes.

(677C) for 4130 steel—the indicator melts and runs. Another application is to use it for repairing high-strength steels used in modern cars. These steels are extremely sensitive to heat. Some can't be heated over 700F (371C) without causing cracking. HP-Books' *Paint & Body Handbook* discusses the repair of automotive high-strength steels. Such information is also available in car-manufacturers' repair manuals.

Avoiding Warpage—Study the following sketches in this chapter and you'll learn to cope with warpage in welding. I said *cope with* because you cannot stop warpage,

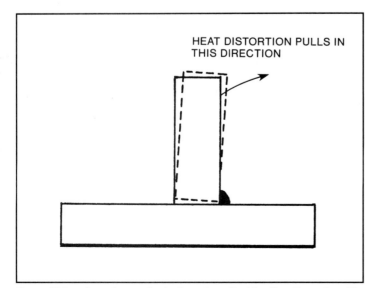

HEAT DISTORTION PULLS IN THIS DIRECTION

As weld bead cools, vertical piece is pulled in direction of weld. Allow for such distortion by leaning vertical section away from side that's to be welded. As weld cools, it will then be pulled toward vertical.

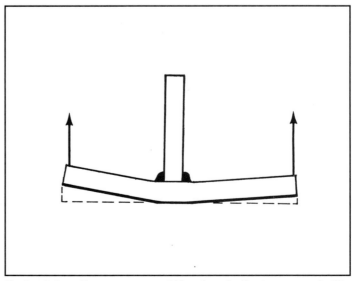

Horizontal section warps as weld bead cools. Such warpage is difficult or impossible to control. To correct it, straighten piece after weld is completed.

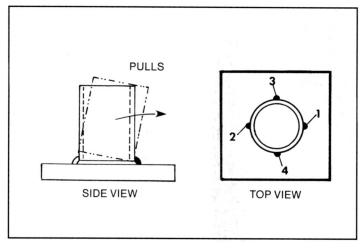

Example of controlling warpage from tack welding: Space tube up from flat plate. Make tack-weld 1, then square tube to plate; make tack-weld 2. Make tack-weld 3, then square tube to plate; make tack-weld 4. Finish weld can now be made.

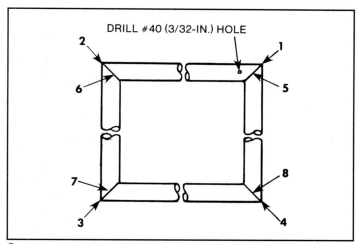

Sequence used to tack-weld tubing in rectangular pattern. Setup is similar to aircraft-fuselage or race-car-frame bulkhead. When welding closed tubing, vent by drilling hole as shown to prevent weld from blowing out as it's finished.

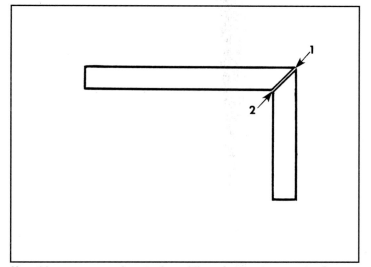

Use this sequence when tack-welding pieces at an angle. Space pieces slightly at weld seam. Make tack-weld 1, then straighten. Make tack-weld 2, then finish weld.

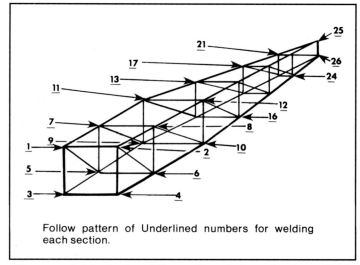

Follow pattern of Underlined numbers for welding each section.

After completely tack-welding entire airplane fuselage, finish welding joints by starting at front where cross section is largest. Whole joint can't be welded at once, so weld *all you can*, then draw torch away *slowly* so weld doesn't crack. Now, go to symmetrically opposite joint and weld in same manner. Later, finish welding joints as you did the first parts. Check alignment often!

just limit it. But you can control warpage.

When building a large tubular structure such as an airplane fuselage, I usually start welding at the front and work toward the back, alternating from side to side. See the above sketch. Even better would be to have two people welding symmetrically opposite sides simultaneously, but that is not done easily. So do the next best thing when welding a large structure: Weld one joint, then the opposite one to cancel the effects of warpage from welding the first joint.

Check alignment after each pair of welds. Repeat this welding-and-checking process until the structure is completely welded. It would be a shame to get a frame or fuselage 80% complete and discover that it's 1/2-in. out of square. A frame, fuselage or large assembly that far out of square is scrap.

Tack-Weld First—Almost every structure should be *tack-welded* prior to finish welding. As mentioned, tack welds are a series of small welds between two adjacent pieces. Spaced about 1-1/2-in. apart, they serve to align the two

pieces, hold them together and help prevent warpage. When the final bead is made, the tack welds are remelted and become a part of it.

Only where the designer calls for complete welding of a joint before welding another section should you bypass the tack-welding rule.

MAKE A WELDING FIXTURE

Much is said about welding *out of position*. You've probably seen bumper stickers noting the various positions weldors are capable of performing in. Although position

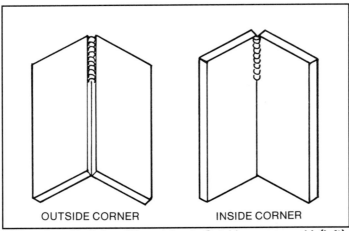

Different welds take different heat. Outside-corner weld (left) takes less heat than inside-corner weld (right).

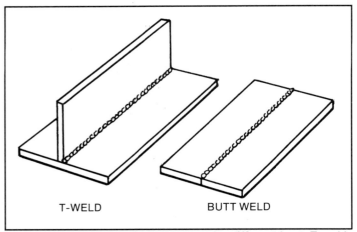

Different weld-joint configurations require different heat. T-weld (left) requires more heat than butt weld (right). Also, majority of heat must be directed toward piece with most mass. For example, majority of heat must be directed toward horizontal piece when doing T-weld.

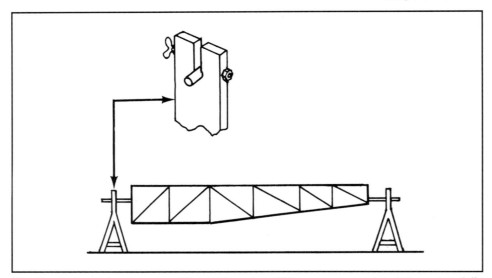

Simple rotating fixture brings weld seam to weldor. Using one saves a lot of time, especially when working alone.

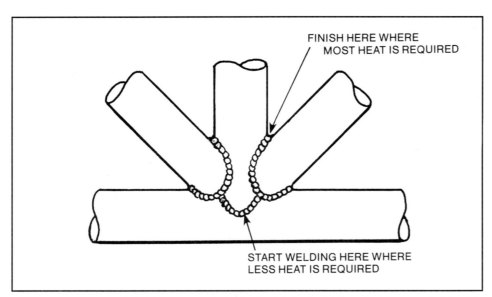

FINISH HERE WHERE MOST HEAT IS REQUIRED

START WELDING HERE WHERE LESS HEAT IS REQUIRED

When welding a complex joint such as this cluster joint, start where least amount of heat is needed. As you move toward area needing the most heat, joint will be preheated, making welding easier and reducing warpage.

in welding terms means the position of the weld surface—flat, vertical or overhead—all experts agree that the weldor should be in as comfortable a position as possible. So, avoid welding while laying on your back or standing on your head if there's an easier, more comfortable position.

The way to get in the best welding position—for both you and the electrode—is to make a fixture for rotating the assembly. If you're welding an airplane fuselage, make a fixture such as that shown at left. Similar fixtures can be fabricated for welding other assemblies. With this fixture, you have better access to all welds. If you're welding a large trailer, hoist it up on its side and turn it over to get to the other side of the weld joints. The "fixture" doesn't have to be exotic. The secret is, make the work accessible to you.

Gas-Welding Equipment

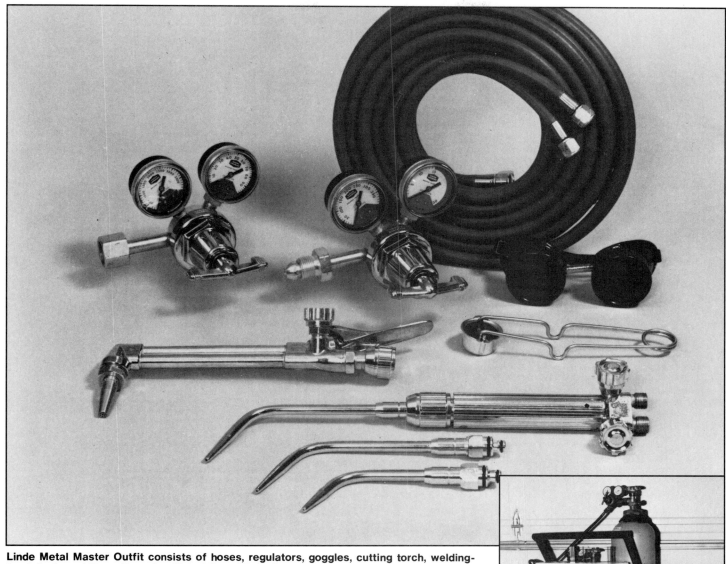

5

Linde Metal Master Outfit consists of hoses, regulators, goggles, cutting torch, welding-torch body and three tips, plus flint striker. All that's needed to oxyacetylene weld or cut are gas cylinders. Photo courtesy of Linde Welding Products.

This chapter describes the equipment you'll need for gas welding—both oxyacetylene and oxyhydrogen. You can be welding before the day is over if you want. But first, you must determine your welding needs. Ask yourself exactly what kinds of things you'll be welding. But, regardless of the type of welding you'll be doing, it's important to master gas welding. Once you've learned to choose the proper tip and adjust the flame, control the heat at the weld puddle and add filler

material, other welding techniques are much easier to learn. Before jumping into the specifics of gas welding, let's first take a look at the basics.

GAS-WELDING BASICS

Energy—heat energy, to be specific—is supplied by two gases, stored in high-pressure cylinders. These gases—acetylene or hydrogen fuel and oxygen—are mixed in specific proportions and ignited to produce a flame temperature of 6300F (3482C) or 4000F

Complete portable oxyacetylene-welding rig includes toolbox for extra torch tips and accessories, tubes for storing welding rod, and hooks for hanging goggles, cylinder caps and wire brush.

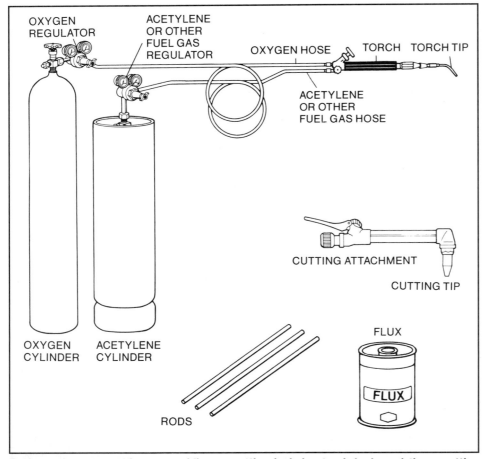

Equipment necessary for gas welding or cutting includes torch body and tip or cutting attachment, hoses, oxygen and acetylene cylinders and regulators. You'll also need filler rod for welding, and flux for brazing or soldering, Chapter 12. Drawing courtesy of Linde Welding Products.

(2204C), respectively. This is more than hot enough to melt steel at 2700F (1482C). To correctly mix these gases and control the flame, special equipment is used: a pressure regulator for the oxygen tank and one for the acetylene or hydrogen tank. A siamesed pair of hoses about 25-ft long carry the separate gases from the regulators to the torch. The torch contains a mixing chamber and two valves to regulate and adjust the flow of each gas, and a changeable tip to obtain the desired flame size and pattern.

After the gases are adjusted for correct working pressure and flow, they are lit at the torch tip. Various flame sizes are produced, depending on tip size. The flame has a dual purpose: to melt the metal to be welded, and to provide a protective gas shield over the molten metal so that atmospheric gases will not contaminate the weld puddle before it solidifies. The torch can be manipulated as necessary to melt the base metal. By adding filler rod, two pieces of metal can be welded together.

GAS-WELDING TORCH

A gas-welding torch consists of a torch body, inlet valves and a removable outlet tip. The torch body includes a mixing chamber for oxygen and acetylene. Valves at the torch inlet regulate flow of the two gases. At the outlet end, various-size tips can be fitted to obtain the desired flame size and pattern.

Torch Size—If you plan to build a homebuilt airplane or formula-type race car, you'll need a small, high-quality gas-welding torch. But if your plans include building a racing stock car or farm trailer, you'll need to supplement that small torch with an arc welder capable of welding heavy-gage material.

I don't recommend a *large* welding torch because it is just too cumbersome to use. And weld quality will suffer if your concentration is broken by muscle fatigue from manipulating a heavy torch. Instead of using a large gas welder, use an electric welder if you plan to weld steel, aluminum or other metals more than 3/16-in. thick.

Compare welding torches to hammers. A 10-lb sledge is effective for driving in fence posts, but you wouldn't want to use it for driving nails. A lightweight hammer lets you work longer without getting tired, and you'll be able to do more-accurate work. The same applies to a welding torch. A small one allows you to make more accurate welds and is less tiresome to hold. That makes a big difference when you're making hundreds of welds on an airplane fuselage or numerous ornamental objects.

TORCH TIPS

The chart, page 38, shows the relationship between welding-tip numbers and tip size for various brands. As you can see, tip numbers are not necessarily descriptive of orifice sizes or the amount of heat a tip can produce. Although the tip number is stamped onto each tip, welding-equipment manufacturers have different numbering systems. There is no standardization.

If you are unfamiliar with a particular manufacturer's numbering system, check tip-orifice size with the shank end of a number drill. Drill sizes run from #1 through #80. For example, a #1 drill fits a 0.228-in. orifice—that would be a "monster" tip. Likewise, a #80 drill fits a 0.0135-in. orifice. If you don't have a chart referencing drill numbers to actual size, mike the

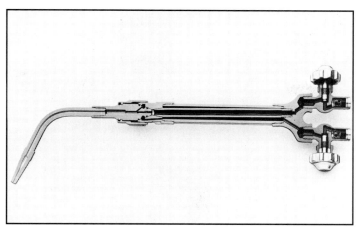

Cutaway of gas-welding torch body and tip. Note mixing chamber in torch tip. Photo courtesy of Victor Equipment Co.

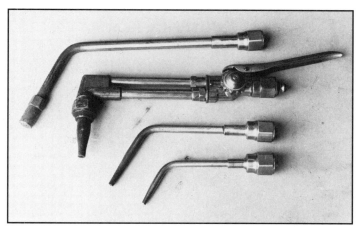

From top to bottom; rosebud tip for heating, cutting head and tip, and two welding tips. Tips may look similar, but will not interchange between brand names. Photo by Tom Monroe.

Easy way to compare prices and features of oxyacetylene welders is by studying mail-order catalogs.

Gas-welding-and-cutting outfit displayed at a Montgomery Ward store: Benefit of this is being able to physically inspect merchandise before spending your money.

shank end of the drill. Never force a drill through a tip orifice—you'll damage the tip.

For most oxyacetylene welding, I use a welding tip with a 0.040-in. orifice. For light work, such as thin-wall 0.020-in. aircraft tubing, I use a 0.032-in. tip. However, if the metal is so thick that I would need a 0.050-in.-or-larger tip, I use my arc welder instead.

How do I know if a tip is too small? Here's a good rule of thumb. If it takes longer than about ten seconds to make a weld puddle at *standard pressure and flow*—see page 38—the *tip is too small* for the metal thickness. It should take about five seconds to make a puddle under normal conditions.

Within limits, you can adjust oxygen and acetylene flow at the

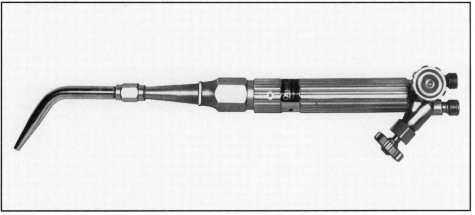

Linde Oxweld torch is capable of welding 0.031—1.0-in.-thick steel. Torch is used in heavy-equipment repair shops. Photo courtesy of Linde Welding Products.

Welding Tips	Hole Size (in.)	Craftsman Tip No.	Victor Tip No.	Smith's Tip No.	Metal Thickness (in.)
	0.020		#000		0.020
	0.025		# 00		0.025
	0.032			AW203	0.040
	0.035		# 0		0.050
	0.040		# 1		0.060
	0.042	#3		AW205	0.090
	0.045		# 2		0.125
	0.050	#5	# 3		0.150
	0.070	#7	# 5		0.250
Cutting Tips	0.018		6 Outer Holes #0-5-101	6 Outer Holes #MC12-10	0.090 to 0.250
	0.020		6 Outer Holes 00-5-10		0.250 to 0.375
	0.025	6 Outer Holes #1			0.375 to 0.500
	0.032		Center Hole #00-5-10		0.500 to 1.000
	0.035		Center Hole #0-5-101		1.000 to 1.500
	0.038			Center Hole #MC12-0	1.500 to 1.750
	0.046	Center Hole #1			1.750 to 2.500
Rosebud Tips	0.042		6 Holes #6-MFA-J		

WELDING, CUTTING & HEATING-TIP APPLICATIONS

Refer to chart when selecting tip for cutting or welding.

Two-stage oxygen regulator maintains constant line pressure as cylinder pressure drops during use. Cylinder pressure of 2000 psi is regulated to 2—35 psi for normal gas-welding or cutting operations. Photo courtesy of Linde Welding Products.

tip to get more or less heat, thereby eliminating the need to change tip sizes with metal-thickness changes.

In most workshops, four different tip sizes will cover the majority of gas-welding jobs: 0.020 in., 0.032 in., 0.035 in. and 0.042 in. You can buy other welding tips, but you'll probably never need them.

Standard Pressure & Flow—The most common error that beginning and self-taught weldors make is using *excess pressure* at the torch. *Maximum* pressure supplied to the torch should not exceed 1 psi for each 0.010 in. of tip opening. For instance, 2-1/2-psi oxygen and acetylene pressure should not be exceeded with a 0.025-in. tip. Be particularly careful about using excessive oxygen pressure. *Never* try to weld with 20-psi oxygen. That much oxygen pressure will oxidize the weld and cause crystallization and cracking of the base metal. See the chart, page 46, for *recommended* pressure.

PRESSURE REGULATORS & GAGES

Simply put, a *pressure regulator* is a mechanical device for delivering a gas to the torch at a *constant reduced pressure*—regardless of how much *higher* supply pressure is at the tank. It consists of an orifice controlled by a spring-loaded valve seat. Spring force and, therefore, gas *flow* is adjustable by a threaded T-handle or knob. A diaphragm seals air pressure against gas pressure. One regulator is required for each cylinder.

Gages—On each regulator is a pair of pressure gages. One reads cylinder, or high pressure; the other indicates line, or low pressure. Line pressure is the pressure supplied to the torch.

The oxygen-cylinder high-pressure-gage range is 0—4000 psi; low-pressure-gage range is 0—150 psi. Acetylene-cylinder high-pressure-gage range is 0—400 psi; its low-pressure-gage range is 0—30 psi with a 15-psi *red line*. This 15-psi pressure *must not be exceeded*. Otherwise, the acetone stabilizer could be drawn from the acetylene cylinder, rendering the cylinder contents unstable and potentially explosive.

Never remove or replace pressure-regulator gages. Instead, if a gage must be removed for any reason, the regulator should be repaired by a welding-supply repair shop. Think of the regulator and gages as precision instruments, with extremely delicate calibration.

Regulators are available in two types: *single-stage* and *two-stage*.

Single-stage regulator: Tightening brass adjusting screw compresses adjusting spring which, in turn, increases line pressure. Photo courtesy of Victor Equipment Co.

Two-stage regulator: Gas-cylinder pressure is first reduced by spring and diaphragm at bottom, then by adjustable spring and diaphragm at top. First stage lowers high pressure to value more controllable by second stage. Note sintered-metal filter in regulator inlet, right. Photo courtesy of Victor Equipment Co.

Single-stage oxygen regulator: Never switch oxygen and acetylene regulators. Different pressure capacities may cause a spontaneous explosion! Photo courtesy of Linde Welding Products.

Light-duty oxygen regulator with 1-1/2-in.-diameter gages is OK for occasional welding. Photo courtesy of Linde Welding Products.

Single-stage regulator drops cylinder pressures from up to 2200 psi to 2—3 psi in one stage. Most small gas-torch kits come with single-stage regulators. The biggest problem with single-stage regulators is that they allow outlet pressure to drop as inlet pressure drops. Also, pressure changes with temperature. High temperature raises pressure and vice versa. Therefore, you must keep your eye on the regulator gages to maintain the desired outlet pressure.

Two-stage regulator automatically reduces cylinder pressure of 2200 psi down to about 30 psi. Pressure is adjustable down to 1—15 psi. Many large gas torches come with two-stage regulators because they use gas much faster—cylinder pressures are likely to drop rapidly due to a high gas-flow rate.

Although it would be nice if a small torch were available with two-stage regulators, many race cars or certified airplanes have been welded with small torches fed by single-stage regulators. You can always buy a set of two-stage regulators later if you absolutely must have the best equipment.

Regulators represent about 75% of the cost of a complete gas-welding starter kit. This doesn't include the cost of the cylinders.

HOSES

Specially bonded, three-layer hoses are required to carry low-pressure gases from the gas cylinders to the torch. The two hoses are siamesed to prevent tangling. Acetylene hoses are red and have left-hand threaded fittings on both ends—identified by a groove around the middle of each brass nut. Oxygen hoses are usually green—sometimes black—and have right-hand fittings. The color-coding and thread differences prevent hose mixups.

Most hoses that come with starter sets are 12—15-ft long; OK for beginners, but too short for serious use. If you have these short hoses, get a set of couplings and extra hoses to make your hoses 25—30-ft long. Longer hoses help you avoid moving the cylinders while working on most welding projects.

ACCESSORIES

Even though you may plan to buy a gas-welding kit with several

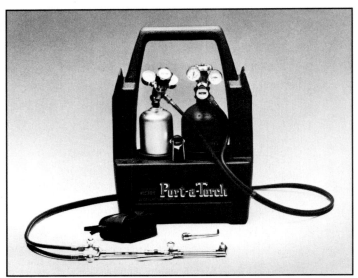

Victor Port-a-Torch is ideal for a home workshop. It is light, compact and keeps equipment organized. Handy carrying handle makes it easy to use at racetrack or for emergency repairs. Photo courtesy of Victor Equipment Co.

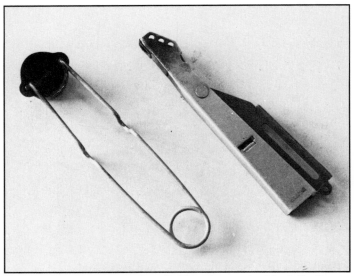

Old standby flint-type striker (left) and new electrical-discharge type (right). Never light your torch with an open flame, particularly from a butane cigarette lighter. You could blow off a hand! Photo by Tom Monroe.

Smith's Cavalier gas-welding kit contains most equipment needed to start gas-welding aircraft. Kit also includes hoses.

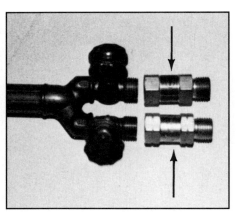

If your torch doesn't have safety check valves, or flashback arrestors, install one between torch and each hose (arrows).

Smith's aircraft-torch oxygen regulator features built-in safety check valve where hoses connect to regulators. Acetylene check valve is shown (arrow).

welding tips, you'll still need a few extra things to make welding easier.

Torch Lighter—Because you must **never light a torch with an open flame,** get a torch lighter, or *striker.* Flint-type lighters make sparks similar to the way you would strike a match, and will work until the flint striker wears down. More expensive, electrical-discharge torch lighters will last many years.

Long-Handle Wire Brush—Use this brush to clean rust and welding scale from parts *before* welding. Weld seams must be clean and rust-free for complete fusion of base metal and filler metal.

Stainless-Steel Wire Brush—Use this small brush to clean welding scale *while* welding. I keep such a brush in my back pocket so it will be handy for cleaning off scale that develops during rest periods.

Welding-Cylinder Wrench—This wrench is handy for removing and replacing the pressure regulators while changing cylinders. Also, this wrench can be used for opening and closing the acetylene- or hydrogen-cylinder valve if it doesn't have its own knob.

Pliers—Use pliers to pick up hot pieces of metal just welded. You can also use them to hold pieces in place while you tack-weld them.

Small Machinist Hammer—Often, you'll need a small hammer to bend hot metal into place, or tap a part into place before continuing the weld.

Safety Glasses—These clear-lens glasses are an absolute must for a weldor's toolbox. Don't confuse them with welding goggles. Wear safety glasses to protect your eyes whenever chipping, filing, grinding or sawing metal. Welding goggles reduce available light too much.

Leather Gloves—Wear leather gloves to shield your hands from welding heat. They allow you to weld for longer periods of time without the need to stop to cool your hands. Because they'll burn, never pick up hot metal with leather gloves. Use pliers instead.

Soapstone Marker—This chalk-like marker doesn't burn off until the metal melts. Use it for making reference marks on metal or to mark lines for a cutting torch.

Temperature Indicators—Temperature-indicating crayons and paint are convenient for determining the temperature of metal for heating or forming. Read about these on page 18.

Welding-Tip Cleaner—As with paint brushes, gas-welding and gas-cutting tips must be cleaned. Cleaning the outside of a tip is easy, but cleaning the inside requires a spiral tip-cleaning *rod*. The cleaner comes with a variety of rod sizes, each matched to a specific tip size. Each rod has a precision fit in its tip hole. Use the tip cleaner as you would a rifle-bore rod.

Metal Files—I use three different files in my welding toolbox: a coarse round file, a coarse half-round file and a flat mill file. Files are used to fit parts before welding.

Acetylene-Regulator Adapter—You might get an exchange acetylene bottle with *male* threads. Consequently, the male-thread (standard) regulator won't screw on. To avoid this problem, get a male-to-male-thread adapter for acetylene, shown in photo at right.

Welding Cart—Instead of buying a welding cart, build your own. It's relatively easy and you'll gain valuable experience doing it. I built mine many years ago and it's still in use. You need training and practice before you start a complicated welding project. So, what better way to gain experience and proficiency than to build something simple, but useful? Welding-cart plans are provided on page 135.

To make the cart, you'll need to use many of the procedures described in this book: fitting, cutting, butt welding, corner

Most gas-welding kits do not include tip cleaners, so you'll have to buy one separately. Clean tips are necessary for making good welds. Photo by Tom Monroe.

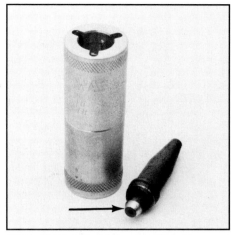

Weld-Aid Products' Re-Seat/Re-Face tool can be used to restore an otherwise unsalvageable cutting-torch tip. Tip self-centers in tool, giving it an accurate reseating job (arrow). Photo by Tom Monroe.

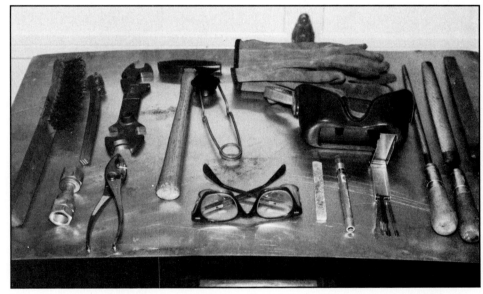

I keep these tools in my gas-welding toolbox. From left to right: wire brush, stainless-steel wire brush, acetylene-cylinder adapter, pliers, gas-cylinder wrench, machinist hammer, flint striker, safety glasses, lightweight leather gloves, single-lens goggles, soapstone, temperature-indicating crayon, tip cleaners, rattail file, half-round file and flat file.

welding, T-welding and brazing. Read the sections that apply to each.

WHERE TO BUY A GAS WELDER

After deciding what size gas welder to buy, you need to decide where to purchase it. One factor influencing your decision may be financing. If you want to use credit, try catalog-sales outlets or their retail outlets.

Another consideration is the service you may or *may not* get

after the sale. To check this out, ask for extra welding tips and a tip cleaner for the welder. If they can't sell you these, chances are they won't be able to help you with replacement parts later on. Try another store.

Shop several local welding-equipment stores. Being able to inspect the merchandise before you buy is important. Fortunately, the folks at these stores usually are familiar with the products they sell—they aren't part-timers working at minimum wage. Never-

Bert Snedden of Hopper, Inc., Santa Maria, California, explains terms of a 99-year gas-cylinder lease to Jon Linke. Ask salesman to show you the different size cylinders available. Here Bert shows Jon 80-cu ft oxygen cylinder.

Small oxygen and acetylene cylinders are OK for small welding shops. Oxygen cylinder at left has 80-cu ft capacity; acetylene cylinder at right is 115 cu ft.

GAS CYLINDERS

Gas cylinders are also called *tanks* or *bottles*. You'll need two cylinders—one for oxygen and one for acetylene or hydrogen gas. Sizing and where to obtain gas theless, you don't have to buy the torch at the same place you buy, rent or lease cylinders.

cylinders are the biggest considerations in putting together a gas-welding outfit. There are many variables concerning how the dealer sells, rents or leases cylinders. Read about these on page 43.

First, decide which size cylinders you need. In HPBooks' *Metal Fabricator's Handbook,* Ron Four-

nier says that gas cylinders come like soft drinks—in three sizes: small, medium and large.

I prefer using medium-size cylinders. One filling lasts a year or longer, even if I'm doing a big project. I can build a complete race car with one filling. Afterward, there should be enough gas left to fix the kids' bicycles, then do some muffler repairs on the family cars. Medium-size cylinders are easy to handle and don't get in the way like the large ones. Small cylinders must be refilled frequently, which is inconvenient.

Interstate Commerce Commission (ICC) regulations govern the leasing and sale of gas cylinders. To compound this, each state has its own regulations. In many states, large cylinders cannot be purchased, only leased.

Cylinder-Size Codes—There are at least two ways cylinder sizes are specified by welding-equipment manufacturers. One system uses letters such as Q, S, H and W. Unless you memorize the size equivalents, this method doesn't mean much. The other system uses capacity designations such as 80, 150, 275 cu ft, and so on. This second system doesn't mean a whole lot more. So, I've developed a helpful, but unofficial, chart to help you decide what to ask for when leasing or buying gas cylinders, page 43.

What's Inside?—Inside a hollow oxygen cylinder, about 80 cu ft of oxygen at atmospheric pressure (14.7 psi) is compressed into 0.7 cu ft of space. Filled oxygen-cylinder pressure is 2000—2200 psi.

Acetylene cylinders are filled with a porous, calcium-silicate filler. Each time they're refilled, acetone is added to *stabilize* the acetylene gas. Filled acetylene-cylinder pressure is about 325 psi.

Hydrogen, helium, nitrogen, argon and carbon-dioxide cylinders are hollow and thick walled. As with oxygen cylinders, filled pressure for these gases is 2200 psi.

It's difficult to judge how much gas is inside an acetylene cylinder by reading the pressure gage. Pressure varies with temperature. For example, the pressure gage indicates more pressure at 100F (38C) than at 30F (-1C). That's the bad news. The good news is you can determine how much acetylene remains in a cylinder by weighing it. Subtract the empty weight in pounds—it's stamped on the side of the cylinder—from the current weight. Multiply the difference by 14.7. It takes 14.7 cu ft of acetylene to make 1 pound at standard temperature and pressure—70F (21C) and 14.7 psi. The answer is in cubic feet of acetylene. For example, if the cylinder weighs 45-lb empty and now weighs 50 lb, you have 5 lb of acetylene, or 5 lb X 14.7 cu ft/lb = 73.5 cu ft.

This method can be used for other fuel gases. Just substitute the volume per pound of gas from the accompanying chart.

OXYGEN-ACETYLENE CYLINDER SIZES

OXYGEN CYLINDERS

Size	Capacity	Letter Size	Diameter X Height (Approx.)
SMALL	80 cu ft	Q	7 X 27 in.
MEDIUM	150 cu ft	S	7 X 47 in.
LARGE	275 cu ft	H	9 X 56 in.

ACETYLENE CYLINDERS

Size	Capacity	Letter Size	Diameter X Height (Approx.)
SMALL	115 cu ft	S	6 X 24 in.
MEDIUM	140 cu ft	4	8 X 38 in.
LARGE			12 X 46 in.

If partially filled cylinder is returned, don't expect a credit for the unused contents. These sizes vary from one manufacturer to another.

WEIGHT & VOLUME FOR FUEL GASES

Gas	Volume (cu ft/lb)
Acetylene	14.7
Propane	8.65
Natural Gas	24.15
MPS Gas	8.65

Weighing an oxygen, helium, hydrogen, argon, nitrogen or carbon-dioxide cylinder tells you nothing because they are lighter than air. However, the cylinder is near empty when pressure drops rapidly. Some oxygen pressure gages indicate how much oxygen remains.

Buying vs. Leasing—Most welding-supply businesses in the U.S. rent or lease gas cylinders and will refill them. Some sell the cylinders outright. But most businesses prefer to lease or rent them. The rationale used for leasing rather than selling gas cylinders is based on safety. They check the cylinders for maximum pressure through *hydro-static testing*—over-pressurizing them with water. If a cylinder fails, a harmless water leak results as opposed to the catastrophic explosion that would result from pressurized gas.

In reality, leasing makes welding-equipment suppliers more money. The chances of getting repeat business is much greater than if the customer owns the gas cylinders. However, I prefer to own the cylinders so I have more control over the situation. A good rule of thumb is this: If you plan to lease cylinders for more than three years, it's less expensive to buy your own.

Lease Provisions—Gas-cylinder lease terms can run 5, 10, 25 or 99 years. Why only 99 years? Despite the fact that most people rarely live 99 years, there's an ICC law that forbids leasing cylinders for longer periods. It has to do with the mandatory pressure testing. If you decide to lease cylinders, here are some of the standard provisions:

- Cylinder valves are serviced free as long as the cylinders are under lease. In time, valves may develop leaks or become difficult to operate. When you lease, these repairs are taken care of by the welding-supply company without cost to you.
- No hydro-testing is required if the company you leased them from is still in business. Hydro-testing is one great way for welding-gas suppliers to make extra money from those who own their cylinders. Every time I take my cylinders to a new town and new gas supplier, they hit me with a hydro-test fee. Usually, it's about $10.
- No *exchange fee* as long as that company is still in business. An exchange fee is a charge for returning cylinders before their lease expires.
- You can cancel your lease at any time and get a refund on the cylinders. You might want to do this if you move to another area or give up welding. Refunds are based on the following typical schedule:
    75% refund the first year.
    65% refund the second year.
    55% refund the third year.
    50% refund the fourth year through lease expiration.
- Of course, you can transfer your lease to someone else for whatever price you agree on.

MPS GAS
Methylacetylene propadiene stabilized, or MPS, gas is less expensive than acetylene and will do a good cutting job. And, overall heat distribution of the flame is more uniform than acetylene. However, it requires about twice as much oxygen to support the same flame temperature. I would not use MPS gas for welding airplane-fuselage parts. I trust the old standby, acetylene.

WHERE TO BUY
OR LEASE GAS CYLINDERS

You should buy, rent or lease gas cylinders from the supplier that will give you good service *after the sale*. But, this is easier said than done. In most cases, you won't discover the quality of service until a year or two afterward. There may come a day when you need your oxygen and acetylene cylinders refilled for a weekend project.

If you get the cold shoulder where you bought your welding equipment, you might even have to rent or lease *another* oxygen or acetylene cylinder because you can't get your's refilled.

Here are typical problems you may encounter when trying to get oxygen and acetylene cylinders refilled:

● Some small welding-supply outlets do not fill their own cylinders. They send them out once every 2—4 weeks to have them filled. So, if you don't plan ahead, you may have to wait 2—4 weeks before your cylinder(s) gets filled so you can start or continue a project.

● If you don't have a purchase receipt, lease agreement, or bill of sale listing your oxygen and acetylene cylinders by serial number, some welding-supply stores may try to confiscate your cylinders on the pretense that you don't own them. Make sure you have proof of ownership with you, even if you have a long-term lease.

● Some welding-supply shops will simply offer to exchange your empty cylinders for their full ones. That seems easy, except when it comes time for refills. Then you have oxygen and acetylene cylinders with no proof of ownership. In this case, someone could accuse you of possessing stolen cylinders! *Make sure you get a new receipt, showing serial numbers for the exchanged cylinders.*

Note: The gases inside the cylinders cost extra and are seldom credited when you return partially used cylinders.

Here is how to get your oxygen and acetylene cylinders refilled with the least amount of trouble:

● Keep proof of ownership papers in your possession, listing serial numbers, size, brand name and latest hydro-test dates. Some welding-supply dealers won't refill a cylinder that's not currently tested for strength and integrity.

● Call welding-supply businesses in your area and ask about their cylinder-refilling procedure. Take your empty cylinders to the business that seems most cooperative.

● Don't wait until the week you need your cylinders refilled to start looking for a place to get them refilled. Start looking at least a month ahead of time.

● Once you decide on a certain business to deal with, get a receipt when you leave the cylinders for refilling. I once had an empty oxygen cylinder stolen from the welding-supply dealer. It was later found in the weeds near the city waste-disposal site. My name was painted on the cylinder and the man who found it called me. Evidently, the perpetrator decided it could not be used because it was empty and he knew he would be caught if he tried to exchange it.

● Paint your name, phone number and address on your oxygen and acetylene cylinders.

CAUTION: Never scribe or etch any line or mark on a high-pressure gas cylinder. This could cause a stress-riser, leading to rupture of the cylinder.

● If getting your cylinders refilled proves difficult, save yourself a lot of trouble and temporarily rent or lease cylinders until you find a business to accommodate you.

● Commercial welding shops usually bypass these first five items by leasing cylinders. And lease costs are tax-deductible to businesses. Welding shops simply have the welding-supply truck deliver full cylinders and pick up the empties.

6

Gas Welding, Heating, Forming & Cutting

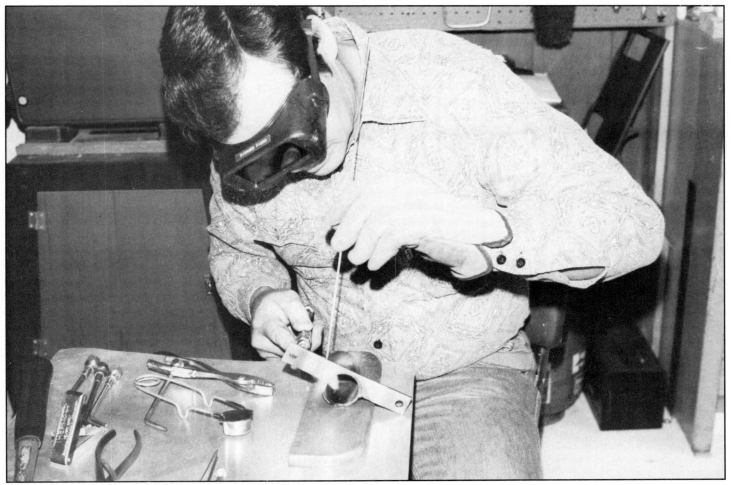

Your work area should be well lit and you should be comfortable while welding. I'm oxyacetylene-welding an airplane wheel-pant chrome-moly tube-and-bracket assembly.

This chapter describes how to gas-weld mild steel, 4130 steel and aluminum. The most important thing to remember is this: When you learn to gas weld correctly, you will have mastered the basics for all types of electric-arc and gas welding! Learning to control the temperature of the molten puddle is the big accomplishment:

Five Basic Steps—You must learn the five basic steps of "metal melting" before beginning a serious welding project.

- Making and maintaining a molten puddle on sheet steel.
- Running a weld bead with a welding rod.
- Butt-welding two pieces.
- Welding an outside corner of two pieces.
- T-welding two pieces.

Once you learn to do these steps, you can practice and practice until you're ready to weld something practical such as the gas-welding cart, page 135.

But, before lighting the torch, get your welding area organized. You need a fireproof platform to weld on. If necessary, you can improvise by using a couple of fire-bricks or a steel-top table, page 139. Remember, there should be *no flammables in the area, and keep a fire extinguisher nearby.*

Read the sections on lighting and shutting off the torch, leak-testing and gas-welding safety *before* practicing.

LIGHTING TORCH

First, rotate the adjusting screws for the oxygen and acetylene regulators counterclockwise one-half to one full turn. This should result in zero pressure at the low-pressure gages when you open the cylinder valves. Too much acetylene pressure could cause problems. Experts contend

SAFETY TIPS FOR GAS WELDING

- *Never tilt an acetylene cylinder on its side when in use.* The acetone stabilizer will flow into the regulator and damage it.
- Mark full cylinders FULL with a marking pen.
- Mark empty cylinders EMPTY with a marking pen.
- Store cylinders at less than 125F (52C).
- Make sure valves are closed and caps are on stored cylinders.
- Chain or otherwise restrain all cylinders in an upright position.
- Always *crack*—slightly open—the valve to blow out dust before attaching a regulator. This helps prevent contamination of the regulator, which may cause erroneous gage readings.
- Never haul cylinders in the closed trunk of a car because of the explosion hazard from escaping gases.
- Wear goggles with the correct filter lens. See the chart, page 13.
- Don't wear oily or greasy clothes when welding.
- Wear leather, wool or denim clothes when welding; leather is best.
- Don't cut or weld material coated with zinc, lead, cadmium or galvanized coating. Poisonous fumes are generated as the coating burns off. Coated sheet steel is used in an increasing percentage of late-model cars and trucks, especially in rust-prone areas such as rocker panels, fenders, rear quarters, door outers, cowl plenums and trunk floors. Read the caution on pages 67—68 for more about this.
- Never use oil or grease on gas-welding equipment. It's extremely flammable at high temperatures, particularly in the presence of oxygen.
- Leak-test hoses, regulators and torch before lighting. Use soap solution or non-oily leak-detector solution on the connectors and look for bubbles. Never use an open flame to check for leaks.
- Oxygen fittings have right-hand threads. Hose is green or black.
- Acetylene fittings have left-hand threads. Grooves around flats on nuts identify left-hand threads. Hose is red.
- Never adjust an acetylene regulator to more than 14 psi. More pressure and flow will draw acetone from the cylinder. Without acetone, acetylene is very unstable and could explode.
- When shutting off the torch, *shut off acetylene or hydrogen first.* Otherwise, you'll soot up the torch and the shop.
- Keep pliers handy to pick up parts you've just welded or cut so you don't burn your gloves or fingers.
- To minimize soot when lighting an acetylene flame, add more acetylene than oxygen. Don't use equal amounts of acetylene and oxygen. The resulting loud pop may cause an accident. Be prepared to add oxygen when the flame ignites. Never try to light the torch with a small amount of acetylene and no oxygen. This causes tremendous amounts of soot.

that if acetylene pressure exceeds 15 psi, acetone will be forced out of the acetylene cylinder, possibly resulting in spontaneous combustion in the acetylene hose.

Next, *fully open* the oxygen valve at the cylinder. Oxygen valves tend to leak unless fully opened or closed.

Now, open the acetylene valve one-half turn at the cylinder. In case of fire, you should be able to shut it off quickly. With half a turn, it is quicker to shut off than with two or three turns. If it has a wrench valve, leave the wrench on the valve.

Adjust the regulators according to tip size. See the chart on this page. For example, if you selected a 0.035-in. welding tip, adjust both regulators to 1-psi pressure. Note that some gages are also calibrated in kilograms/square centimeters (Kg/cm^2).

Open the acetylene valve at the torch about 1/2—3/4 turn and be ready to light the torch with a flint or electric striker. Remember, *never use a match, cigarette or any*

LEAK TESTING

It's always a good idea to leak-test gas-welding equipment from time to time. Do this both after you've set up brand-new equipment or are using equipment with which you're unfamiliar. Why? Aside from the obvious safety considerations of flammable-gas leaks, leakage can cause fluctuating pressure, upsetting torch-mixture settings.

Use a soap solution, such as liquid dish soap. *Never use a bare flame or oil.* Oil is a combustible hydrocarbon, so even a little on the oxygen-hose connection could lead to disaster.

To do a leak test, turn one regulator screw fully *counterclockwise.* Open the cylinder valve. Make sure the torch valves are closed. Build up pressure in the regulator and hose by *slowly* turning in the regulator screw—turn it *clockwise.* This should raise line pressure from about 5 to 15 psi. Now, apply the soap solution to each connection and look for bubbles.

Leak-test both cylinders and their regulators, hoses and the torch.

RECOMMENDED PRESSURES—MILD & 4130 STEEL

Victor Tip Size	Tip Diameter (in.)	Steel Thickness (in.)	Oxygen Pressure (psi)	Acetylene Pressure (psi)
000	0.020	0.020	1	1
00	0.025	0.020	1	1
0	0.035	0.050	1	1
1	0.040	0.060	2	2
2	0.045	0.125	2	2
3	0.050	0.125	3	3
4	0.060	0.150	4	4
5	0.070	0.250	5	5

Adjust oxygen regulator to about 3 psi on low-pressure gage, left. Low-pressure acetylene regulator should also be adjusted to 3 psi.

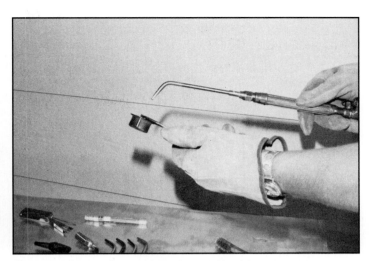

Open torch acetylene valve about 1/2—3/4 turn. Hold striker under torch tip and light torch. Remember, *NEVER LIGHT AN ACETYLENE TORCH WITH AN OPEN FLAME!*

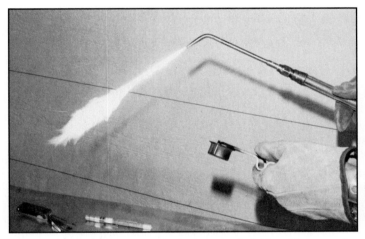

Acetylene-only flame should look like this. Quickly add oxygen to eliminate soot.

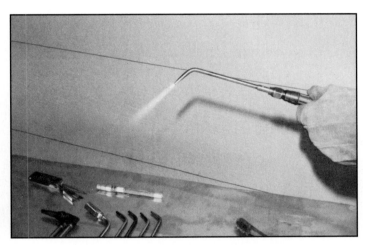

As you add oxygen, five distinct flame cones will appear. Continue adding oxygen until you have a neutral flame. It looks like this with #3 Victor tip, 3-psi oxygen pressure and 3-psi acetylene pressure. One distinct cone is at center with light blue outer flame.

other type of open flame! Now, light the torch.

Flame Types—As you start opening the oxygen valve at the torch, the torch flame should look like the accompanying illustration. This flame is still slightly acetylene-rich, or *carburizing*. There are distinctly separate flame colors in this carburizing flame. Continue to add oxygen at the torch valve until the flame looks like that pictured at bottom right or on the back cover. The outer flame should be about 6-in. long, with a deep blue inner cone about 1/2 to 3/4-in. long, depending on torch size. Just as the middle flame cone disappears, you have the *neutral flame* desired for gas welding.

Practice adjusting the flame a few times until you can adjust for each of the three flames shown, page 48. The *carburizing* and *oxidizing* flames are used for other operations, such as cutting or metalworking, but *not* for gas welding.

A carburizing flame is fuel-rich,

GAS-WELDER WORKING PRESSURES			
PSIG	**KG/CM^2**	**PSIG**	**KG/CM^2**
0	0	225	16
7	0.5	280	20
14	1.0	340	24
22	1.5	400	28
27	2.0	750	50
55	4.0	900	100
115	8.0	2200	150
165	12	2850	200

Pounds per square inch gage (psig) to Kg/Cm^2

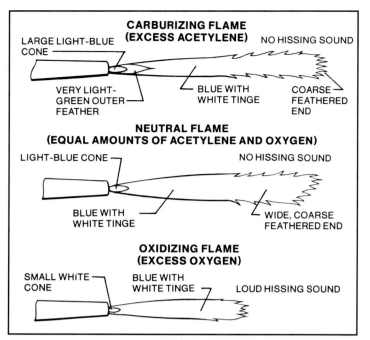

Basic gas-welding flames: Each has distinctive shape, color and sound. Neutral flame is the most used.

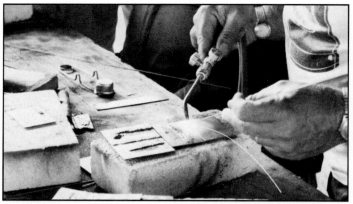

Practice welding on steel plate. Student welder is using firebrick to keep heat off workbench and insulate workpiece.

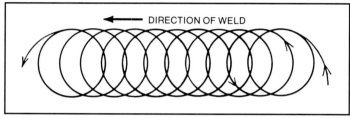

Use circular motion to preheat base metal before forming puddle. Circular motion spreads heat evenly. When steel turns dull red, you are ready to make molten puddle.

as opposed to oxygen-rich. Such a flame is useful in metal-forming aluminum. If a work-hardened area must be annealed, or softened, the area is coated with carbon, or *soot,* using the acetylene-rich flame. Once coated, the torch is readjusted to a neutral flame and the area is heated until the carbon coating burns off. At this point, the aluminum is annealed.

The oxidizing flame doesn't have much use in welding or metalworking. Avoid it.

SHUT OFF TORCH

Always shut off a welding torch when it's not in your hand. More people have been burned and fires started by a lit welding torch propped up against something than I'd care to guess. How you shut off the torch depends on how long it'll be before you reuse it.

If you plan to light the torch again in a few minutes, simply close the torch valves—acetylene first. But if you don't plan to reuse the torch for a longer period, also shut off the cylinder valves after shutting off the torch. With the

cylinder valves shut, there is less chance of a gas leak and subsequent fire or explosion. Even without a fire or explosion, a leaky valve is money down the drain.

CARE OF GAS-WELDING EQUIPMENT

Treat your gas-welding torch and regulators as you would a quality camera, target pistol or any piece of precision equipment. Never let gas-welding equipment get wet or oily. Never leave the torch or hoses lying on the floor where they can be stepped on or driven over. Torch hoses last a long time, but not when subjected to that kind of abuse. And never, ever, lay the torch on the shop floor. Always coil the hoses and hang them on a hook, off the floor.

Do not bump or hit the regulators or gages. These precision-calibrated instruments could be damaged. In most cases, return gas-welding equipment to a dealer for cleaning and repairs—even for torch-tip O-ring replacement. A quality gas-welding outfit should last many years, if properly cared for.

PRACTICE

By now, you should have collected some scraps of steel to practice welding on. Ideal for welding practice would be several 2 X 5-in. pieces of 0.032—0.060-in.-thick mild steel. They don't have to be exactly this size, but your practice work will look better if the pieces are uniform.

GAS-WELDING TECHNIQUE

Light the Torch—If you've selected a piece of scrap steel on which to practice, you have your goggles and proper attire on, and your welding equipment is set up, it's time to light up. Open the valves at both gas cylinders and adjust oxygen pressure to 10 psi and acetylene pressure to about 7 psi.

With the torch lighter in hand, open the torch valves. *Crack* the oxygen valve and open the acetylene valve a little more. Light the torch. Once lit, adjust the torch to a neutral flame. Readjust pressures at the regulators, then at the torch. Pressures tend to change a bit with ambient air temperature. You're now ready to weld.

Making a Puddle—The first thing

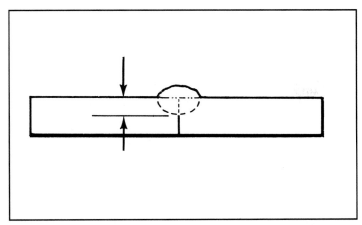

Good weld must have sufficient penetration, or depth of fusion. Weld has 50% penetration. Increase heat or slow down to increase penetration.

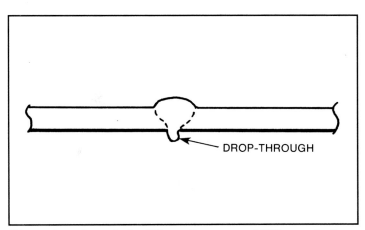

Slight *drop-through* is OK as it ensures 100% penetration. Reduce heat to reduce or eliminate drop-through.

to do is make a molten puddle on the steel plate. With your welding goggles on and over your eyes, direct the neutral flame at the steel. Oscillate the torch tip in a half-moon, zig-zag or circular pattern as shown. The exact pattern is not important. The idea is to keep moving the torch in a rhythmic, repeatable pattern as you move the weld puddle along, but without overheating it. Use the pattern that works best for you.

Working Distance—Hold the torch about 1 in. from the work. The steel should start to turn red within 5—10 seconds. If it doesn't, you're not holding the torch close enough, or the tip you selected isn't large enough for the metal thickness.

I usually tell first-time welding students to overheat the metal first and melt holes in it. This is because the normal tendency is to weld too cold, resulting in poor penetration. So go ahead and burn a few holes until you can control the heat and maintain a puddle.

Learn to *manipulate* the torch by moving it closer to the metal, then backing up quickly if a hole starts to form. After experimenting with the puddle for about 10 minutes, you're ready to advance to the next step.

RUNNING A WELD BEAD

The next step is to practice running a bead on your piece of scrap steel. Start by making another

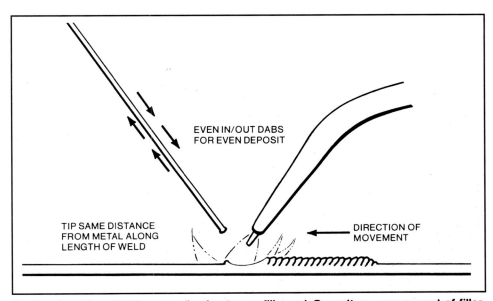

Forehand welding: It takes coordination to use filler rod. Deposit an even amount of filler rod with each dab as you move molten puddle along weld seam. Practice will develop rhythm. Drawing by Ron Fournier.

WELDING WITHOUT FILLER ROD

Most welding textbooks make a big deal about welding without filler rod. But, except for spot welding, welding without filler rod is almost like painting without paint! Welding without a rod is done rarely, unless it's tack-welding one piece to another. Then you make adjacent weld puddles—one on each workpiece—and let them fuse together. Furthermore, a weldor should always have the filler rod in his hand, ready to *fill* the puddle as required to make the weld bead.

molten puddle. With your left hand—if right-handed—momentarily *dip* the welding rod into the

puddle, then withdraw it. If you're left-handed, the torch goes in the left hand and the rod in the right. *Always* dip the rod into the molten puddle. *Never* try to heat the rod and puddle together. If you do, the flame will melt or even vaporize the small-diameter rod before the base metal gets hot enough to puddle. Remember, form a puddle, then intermittently dip the rod into the puddle to add filler material as you run the bead.

You've seen those beautiful welds that look like a row of fish scales? Well, they resulted from a weldor doing the dip, dip, dip thing. If the welding rod sticks in the puddle, point the flame at it, melt it off and try again. Sticking is

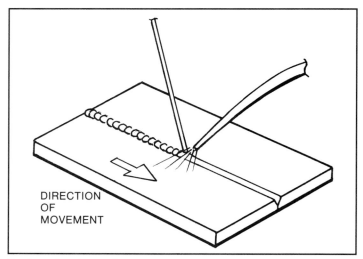

Backhand welding: Seldomly used, weld puddle is moved in opposite direction to that in which torch is pointed.

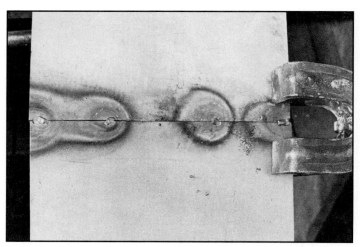

Tack welds hold pieces together for final welding, preheat metal and help prevent warpage. Tack-weld ends of weld seams, then space additional tack welds 1—3-in. apart. Make them closer when welding sheet metal. Note discoloration of heat-affected area. Photo from HPBooks' *Paint & Body Handbook* by Don Taylor and Larry Hofer.

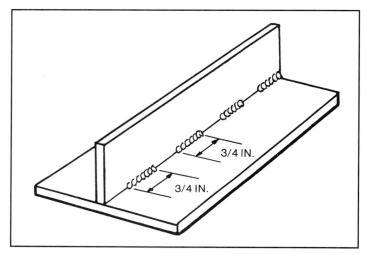

Typical stitch welds: Technique is used where continuous weld is not required for strength.

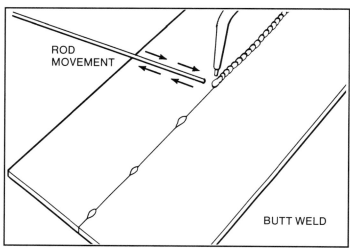

Sheet metal needs plenty of tacks—about every 1 in.—to reduce warpage. Tack welds are melted into weld bead as final bead is made. Drawing by Ron Fournier.

caused by not dipping fast enough or not keeping the puddle molten. Just keep practicing.

Forehand Welding—*Forehand* or *backhand* welding refers to the direction you point the torch tip in relation to the direction you're running the weld bead. If you're forehand welding, the torch is angled so it points in the direction of the weld. This is to preheat the base metal so it puddles easily as you move along with the weld bead.

Backhand Welding—Like walking backward, backhand welding is similar to welding backward. The technique is to point the torch at the already welded seam, away

from the unwelded seam! This prevents the base metal from being preheated—usually an undesirable feature. Backhand welding is used rarely, except to avoid burning through very thin metal. The added mass of weld bead may help absorb the extra heat. However, so does pulling the torch away from the work.

Tack Weld—As previously discussed, a tack weld is nothing more than a very short weld that's used for holding two pieces in place prior to final welding. You'll make a lot of tack welds.

Stitch Weld—A stitch weld is used where a continuous weld bead would be too costly and time-

consuming, and where maximum strength is not required. Although it can vary with the application, a stitch weld typically is made up of short weld beads about 3/4-in. long, spaced by equal gaps.

Butt Weld—Once you've mastered the art of running a bead, you're ready to try welding two pieces of metal together. Let's start with a basic joint. A *butt weld* is a weld made between two pieces laying alongside and butted against one another, edge to edge or end to end.

Place two pieces of metal side by side and butt them together. There should be no gap. The seam will be welded into one solid bead.

To avoid contaminating the weld bead with firebrick or the welding table, raise the metal pieces off the table or brick by inserting extra pieces of scrap underneath each workpiece, but not under the weld seam.

Next, tack-weld the two pieces together, first, at each end, then about 1-in. apart, along the length of the seam. This keeps the metal aligned during welding. The trick here is to keep both edges at the same temperature by *manipulating* the torch. Add a little heat until the puddle forms, then dip the rod in two or three times until you have a good tack weld.

After tack-welding, use your pliers to hold the work so you can check for warpage at the weld seam. Straighten the pieces by tapping on one piece with a hammer when you hold the other with the pliers. You're now ready to run a solid bead. If you're right-handed, do a forehand weld by starting at the right end of the seam, make a puddle—torch in right hand—dip the rod and keep going until you get to the left end. If you're left-handed, reverse hands and start welding at the left end of the seam.

As you come to each tack weld, remelt it into the puddle. When complete, you won't be able to see the tack welds; they'll be part of the weld bead.

Test Weld—Because the appearance of a weld can fool the beginner, test each weld for soundness. "Pretty" welds can literally break in two if there is *insufficient penetration*—not enough filler fused with the base metal. The weld bead may only be "laying" on the base metal. Ideal penetration may be from as low as 15%—weld bead is fused into the base metal by 15% of overall thickness—to over 100%—it's fused the full thickness of the base metal and sagging through to the back side. Clamp the piece to be tested in a vise, just below the weld seam. With a big hammer, bend the top piece *toward* the top of the weld bead. Chances are that

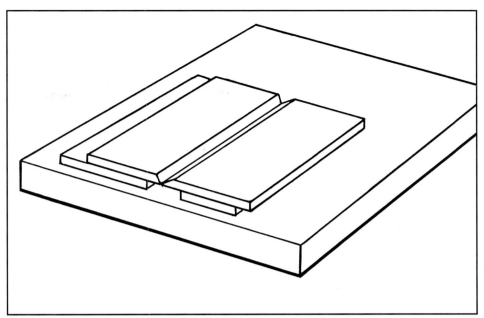

Space pieces to be butt welded off table on scrap pieces or firebrick so you won't contaminate weld, waste heat or damage table top.

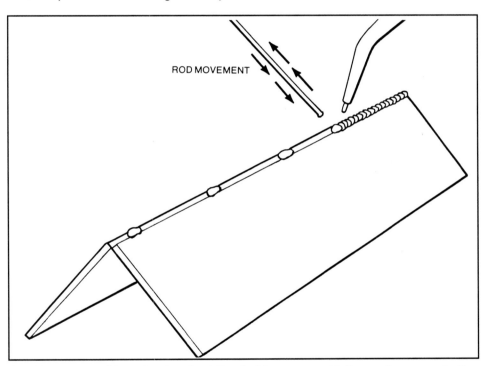

ROD MOVEMENT

Outside-corner weld requires less heat than inside-corner weld. For maximum penetration, metal edges should not overlap, but form a V. Drawing by Ron Fournier.

the weld will break through the back side, perhaps completely if penetration is poor. A common cause of broken welds made by beginning weldors is *crystallization*. Crystallization is caused by excessive gas pressure—usually, too much oxygen pressure.

It is much better to weld with 1—2-psi gas pressure for any size tip and avoid overheating the

weld. If your weld breaks, don't give up. Look for lack of penetration. With your next test weld, try to get a good puddle going before you dip in the filler rod. Practice makes better welds.

Outside-Corner Weld—A fillet, or outside-corner, weld is a weld performed on two pieces of metal joined in a V-type configuration. The weld bead is run on the out-

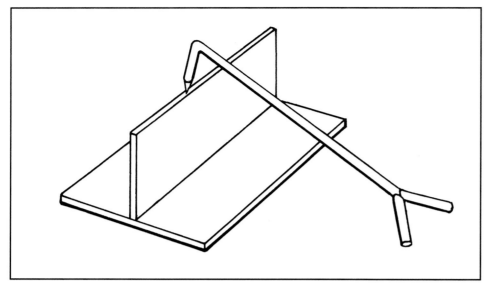

When making T-weld, support vertical piece with mechanical finger. Tack-weld each end of joint. Point flame at horizontal piece 70% of welding time, but *manipulate* torch to put *equal heat* on both pieces as you run bead.

Rosette, or *plug,* weld (arrow) is made through hole. This is done where one piece overlaps another to gain additional strength. Threaded plug is inserted in and welded to thin-wall tube on this race-car supension member. Photo by Tom Monroe.

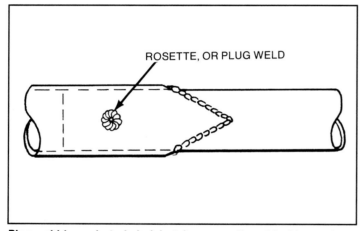

Plug weld is made to help join tubes, one slipped inside another. Note angle-cut—*scarfed*—end of larger tube to increase weld-bead length.

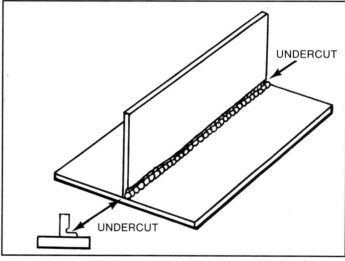

Problem with making T-weld is undercutting vertical workpiece and insufficient penetration on horizontal workpiece. Undercutting can be eliminated by correct manipulation of torch and rod. If your weld looks like this, point flame at undercut and add filler rod as puddle forms there.

side rather than in the V's inside, or "crotch." An outside-corner weld is easier to make than a butt weld because it takes less heat to maintain a puddle and run the weld bead. It's easier to get the edges hot because they're up in the air and only the edges are being heated. Less heat is lost to supporting members, such as the steel work table.

Block the two pieces of metal up like the peaked roof of a house. You can use a heavy piece of metal at the sides to hold the two pieces in place for tack welding. When you do this, you have just created your first welding jig!

Tack-weld the two ends, then make some tack welds in between. Now, run a continuous weld bead. A near-perfect corner weld is one with slight penetration through to the underside—100% penetration of the base metal.

T-Weld—T-welds are the most difficult of all the practice welds. That's why we waited until last to try it. When making a T-weld, you're making another type of

fillet weld—welding in a corner where two pieces join at 90° or so.

Block up two pieces of scrap metal with a metal finger as shown in the accompanying illustration. Viewed from the end, the two pieces form an upside-down T. Block up the flat piece in the area of the weld so the welding table won't absorb the heat of the weld. Firebricks or short sections of angle stock are great for this.

Next, tack-weld the two ends as before. Then make more tack welds about 1-in. apart along the weld seam. Again, if you are right-handed, start welding from the right end—vice versa for you southpaws. Direct most of the heat to the horizontal piece and less to the vertical—at about a 2-to-1 ratio.

The reason for directing more heat at the horizontal piece is that the weld is being made in its center, so there's more volume of metal to absorb heat. There's only half as much volume in the vertical piece because you're welding its edge. Remember, *manipulate* the torch to keep the puddle going on both pieces, and feed the rod into the puddle by intermittent dipping.

A common problem with T-welds is ending up with an otherwise decent weld that has an *undercut* or gouge in the vertical piece. This is caused by failing to adequately manipulate the torch to keep equal puddles on both pieces and overheating the vertical piece. The molten base metal from the vertical piece runs down and solidifies into the weld bead.

The solution is to tilt the torch away from the vertical piece when you see the undercut start, and at the same time, dip the rod into the undercut part of the puddle. You'll have to do a little *torch twisting,* but that's what it takes—that and a lot of practice.

Remember, control the temperature and the puddle, and the weld will take care of itself.

COMMON PROBLEMS

Torch Pop—This will scare the dickens out of you and blow sparks all over the place! *Torch pop* is ignition of the gases *inside* the torch. It is more common when welding in corners such as when making a T-weld. Torch pop is caused by an overheated tip due to holding the torch tip too close to the metal. This causes a small explosion inside the tip. After the torch pops several times, the tip gets *dirty* from the metal splattered on it, causing it to pop even when you aren't too close.

The solution for preventing

torch pop caused by a dirty tip is to shut off the torch and clean the tip with your tip cleaner. Light the torch and try welding again. If cleaning the tip didn't cure the problem, *increase oxygen and acetylene pressures* 1 psi from the previous settings and adjust for a hotter flame. If this doesn't cure the popping problem, try the next *larger* tip.

Flaky Welds/Poor Penetration— Such welds break apart when you bend them. They're caused by not making the puddle hot enough before dipping the rod. You just can't melt rod and drop it onto the base metal, hoping it will stick. Instead, it must become a homogeneous part of the base metal by mixing, or fusing, while in the molten stage.

The solution is to get the weld puddle hotter. Do this by using a larger tip or by holding the torch closer to the work, but not so close that torch pop results. Remember, if you can't make a molten puddle in 5–10 seconds, use a larger tip. This is why I tell new weldors to melt holes in the metal, if necessary, but get it hot! Usually, it takes no more than 15 minutes of extra practice and you can be making good welds with this technique.

Rod Sticks to Base Metal—Every beginner experiences rod-sticking problems because the weld puddle isn't hot enough. The solution is simply to heat and maintain a molten puddle. The *puddle* melts the rod, not the torch. If you keep a good puddle going, then merely dip the rod where you want filler material. You'll get a good weld bead. Concentrate on the puddle and the weld will take care of itself.

Flashback—This is a potentially dangerous condition where the gas burns back through the torch and hose to the regulator and cylinders, damaging the torch, hose and regulator. The cylinders are next and an explosion is possible!

Flashback is usually accompanied by a loud hiss or squeal. If it

occurs, *flashback must not be allowed to continue!* Immediately shut off *oxygen* at the tank if flashback occurs, then shut off acetylene. The oxygen is first because it supports combustion. Flashback is usually caused by a clogged torch barrel or mixture passage. Don't relight the torch until you cure the problem.

For these reasons, every oxyacetylene torch should be equipped with flashback safety arrestors. Basically, these are one-way valves that install in the torch gas lines.

Solutions—Don't be afraid to move the torch as necessary. You have 6300F (3482C) available at the tip of a gas torch. Position the tip close enough to a piece of steel that melts at 2750F (1510C), and the metal *will* melt!

If the puddle gets too big, pull the torch away for a second or two to give the metal a chance to cool and solidify. If you burn holes on one side of the weld seam and are not getting the other side red, direct the torch away from the hot side and toward the cold side to get more-even heat.

You are probably getting tired of hearing this, but *temperature control is the key.* Control the temperature and you control the weld puddle. After selecting the correct tip size, temperature is controlled by the direction of the torch, the distance the tip is from the work and by gas-pressure settings. Follow these simple rules:

- **Point** the torch where you want the heat.
- **Aim** the torch away from where you don't want the heat.
- **Back away** if the puddle is too hot.
- **Move closer** if you are not getting the puddle hot enough.
- **Increase** heat by opening the torch valves if you can't get enough heat.
- **Decrease** heat by closing the torch valves if the puddle is too hot.
- **Move** the torch in an oscillating pattern.

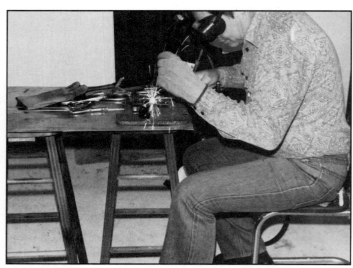

Never weld 4130 steel in a draft or breeze. Thermal shock to 4130 steel may cause cracking at weld seam. At this outdoor demonstration you can actually see sparks being blown by a breeze. The weld is probably ruined.

Note clean floor and lack of flammables in weld area. Temporary weld table consisting of sheet steel on stools works better than kneeling on floor to weld.

GAS-WELDING SAE 4130 STEEL

After you've built your gas-welding cart, and maybe another useful project or two, you should be ready to try welding chrome-moly, or SAE 4130, steel. SAE 4130 welds about the same as mild steel, but it's more likely to become air-hardened and brittle from improper welding. Nevertheless, don't be afraid of 4130. If you can weld a nice bead with mild steel, you can learn to do the same with 4130.

Don't use copper-coated rod for welding 4130 steel. It may cause cracks and bubbles in the weld. Use *only bare mild-steel or bare 4130 rod.* For most jobs, 1/16-in.-diameter rod is the best size to use. It comes in 36-in. lengths. I cut them in half for better control—did you ever try writing with a yard-long pencil?

Never braze 4130 steel. Its wood-like grain will open up and let brass flow into it. When the brass solidifies, the steel will then have thousands of little wedges that cause cracks between the grains. Sometimes the cracks will propagate as you watch!

Cleaning—Keep 4130 tubing or sheet clean of all oil, rust and dust. *Clean it before you weld it.* Don't even touch the weld area with your fingers after cleaning. Use

Main landing-gear strut-and-axle assembly is gas-welded 4130 steel. Axle was fitted through hole in main strut, then welded.

methylethyl ketone (MEK), acetone or alcohol to clean both the base metal and welding rod. You can't get it too clean!

File, sand or sandblast all scale from previous welds before welding over them. The scale could contaminate your weld if not removed.

Shop Area—Keep the welding area clean, well-lit and draft-free —especially for welding 4130. A bright, clean shop area helps you make clean welds. A dark, dirty welding shop will contribute to cracks, pinholes and generally poor welds. Before you actually start building a long-term project,

go out to the welding shop and write someone a letter in the position you'll be welding in! If you're not comfortable writing the letter, you certainly won't be comfortable welding. And never allow any drafts of air, cold or hot. One welding instructor once advised me to not even let my dog wag his tail in the welding shop!

Weld Technique—Evenly preheat the weld area to about 375F (190C). Although preheating to the precise temperature is not critical, a temperature-indicating crayon or paint can be used for getting the feel for how hot this is. *Play the flame*—move it back and forth—over the entire weld seam, holding the torch tip about 4 in. from the metal.

Start welding where a minimum of preheating is required to form a puddle—such as on the edges. After running a bead for a fraction of an inch or so, the metal is automatically preheated, particularly if you're using the forehand method. This saves preheating time and reduces the chance of overheating the weld.

If you tack-welded the seam prior to running the final weld, be sure to remelt the tack welds along with the base metal and include them in the weld as you come to each.

Never jerk the torch away as

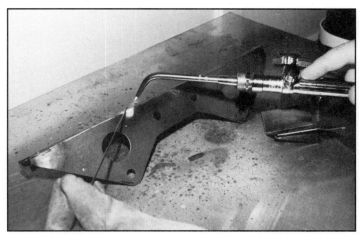

Tack-welding outside-corner weld of 4130-steel engine bracket: Note how torch is held and pointed with right forefinger. As soon as puddle forms, welding rod is dipped into it to complete tack weld.

Homemade tow bar is good project to practice welding 4130 steel. Small wheels at front landing gear are used in manufacturing to prevent flat-spotting or damage to actual wheel and tire.

Nose-gear assembly on Durand biplane is gas-welded 4130 steel.

Uncovered biplane constructed of gas-welded 4130 steel was displayed at Oshkosh.

you complete a weld. Hydrogen and oxygen in the air will contaminate the weld and it will cool too rapidly, possibly cracking it. After finishing a weld, pull the torch back slowly. Let the weld cool to a dull red before removing the torch completely. Not only does this reduce the chance of cracking the weld, pulling the torch back slowly also allows the molecules to relax gradually and *stress-relieve*, page 30, somewhat. Even when stopping for more welding rod, hold the torch 4 in. from the work so the flame "bathes" the weld in heat.

Never weld the back side of a

4130 weld *unless* the designer specified it. If welded properly, the joint will be strong enough without doing so. Besides, the back side of a weld probably has scale that should be sandblasted prior to welding.

When redoing cracked welds on 4130 steel, file or saw out any bad welds and start over. You might even have to put a *patch plate* over the joint if excess metal was removed. A patch plate is usually made of the same material and thickness as that being patched. Extend the patch plate 200% past the damaged area and weld the plate all the way around.

When gas-welded mild-steel airplane exhaust system is completed, it will be sandblasted and painted with ceramic paint.

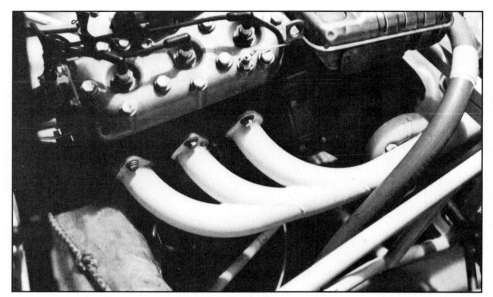

Mild-steel exhaust headers on SAAB 3-cylinder "corn popper" engine were gas-welded. To improve their looks and inhibit rust, they were sandblasted, painted with ceramic-base paint, and oven-baked.

Gas-welding project that will provide fun later is go-kart frame. See page 142 for plans.

Corvair fender and door-skin replacements are to be gas-welded. Note replacement panels tack welded in place.

Oxyhydrogen is optimum setup for gas-welding aluminum. Cylinder at left is oxygen; cylinder at right is hydrogen. Oxygen regulator is adapted to hydrogen cylinder.

Drill a relief hole in tubing that's being welded closed. If you don't, air pressure building up from heat inside the tube will blow out the last of your weld as you finish sealing the tube. Therefore, drill a #40 or 3/32-in. hole in a non-stressed area about 1 in. from the end of every tube to be welded shut. If you want, squirt spray preservative such as LPS-1 or WD-40 into this hole. Or you can leave the tube dry as I recommend, and either weld the hole shut or seal it with a Pop rivet. If you rivet the hole shut, coat the rivet with silicone sealer to keep out moisture.

Rust Prevention—Rarely is it necessary to add oil to preserve the inside of a 4130-steel tubular structure. If moisture can't get inside the tubes, they won't rust. Most rust occurs from outside. Paint will protect it there. Oil is heavy, messy, and may contain chemicals harmful to 4130. I've repaired rusty fuselages from airplanes built in the '30s, and the rust was on the outside, not inside. There was no oil preservative inside the tubing.

AUTO-BODY GAS-WELDING

From the first steel-bodied cars until about 1980, gas-welding was the only way to repair severely damaged car bodies. After 1980, an increasing number of car bodies use high-strength steel (HSS) panels. Depending on the alloy, many high-strength steels have a crystalline grain structure that can be destroyed if heated above a certain temperature. The limit for *martensitic* steel, for example, is 700F (371C). This severely weakens the metal and can cause cracking.

For this reason, auto manufacturers recommend that HSS panels be welded with other techniques—MIG welding is one of them. But for all those older cars, you can make good use of your 6300F (3482C) gas welder for body-and-fender repair.

To determine whether the body panel on a late-model automobile or light-duty truck is HSS, consult the shop manual. Because each manufacturer uses different types of HSS and may have different repair procedures for even the same alloy, follow its procedures to the letter. Many HSS panels are critical structural members and the repair procedures are difficult, so don't be surprised if the manufacturer recommends *replacing* a damaged HSS panel rather than repairing it.

GAS-WELDING ALUMINUM

Most people equate oxyacetylene welding with gas welding. That's because acetylene is by far

Oxyhydrogen flame is almost colorless. Small, 2-in.-long torch body for welding aluminum is light and easy to handle.

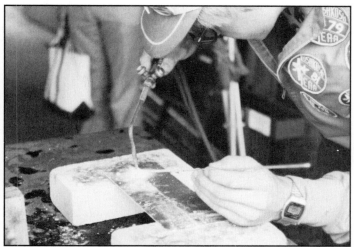

Experimental Aircraft Association (EAA) welding instructor butt-welds 6061-O aluminum sheets. Flux paste makes metal look messy. However, it's necessary to ensure a sound weld.

the most popular fuel used for gas welding. But when it comes to gas-welding aluminum, hydrogen is often recommended instead of acetylene. If you do use hydrogen in place of acetylene for welding aluminum, be sure to switch back to acetylene to weld steel. Welding steel wih oxyhydrogen will cause *hydrogen embrittlement*—hydrogen contamination of the weld joint, causing it to be brittle.

Oxyhydrogen—For many years, aluminum welding was a mystery to me. I thought TIG was the only way to weld aluminum. Then I bought some aluminum welding flux, bare aluminum rod and tried gas-welding aluminum. It surely is *different* from gas-welding steel, but it works!

Oxyhydrogen is the preferred method of welding aluminum. This is because its 4000F (2204C) neutral-flame temperature is closer to the 1217F (658C) melting temperature of aluminum than is the 6300F (3482C) flame temperature of oxyacetylene.

Aluminum vs. Steel—When using oxyhydrogen to weld aluminum, remember that an oxyhydrogen flame has little color. It doesn't look like an oxyacetylene flame. Therefore, don't try to adjust it to the same color. It will have two distinct cones—inner and outer—as described for oxyacetylene welding. When properly adjusted, oxyhydrogen flame has an almost clear outer cone and pale-blue inner cone. However, the adjusting procedure is the same.

Use hydrogen just as you would acetylene. Almost all the procedures for lighting the torch, choosing tip size, tip-to-work distance and other techniques for welding aluminum are essentially the same as for welding steel. Temperature control through practice is still the secret, regardless of the material or method.

One major difference in welding aluminum compared to steel is that the puddle is much cooler than steel and will not melt the welding rod as easily. So you must preheat the welding rod slightly by holding it near the puddle and partly in the flame as you move the puddle along. The dip, dip, dip process is the same.

Another difference with welding aluminum vs. steel is that there's little change in base-metal color as it melts to form a puddle. Aluminum doesn't change color and glow red like steel when it's heated. Instead, it stays the same color until a few degrees before reaching its molten state. At this instant, it becomes shiny where the puddle forms. If you continue heating the aluminum puddle past this state, looking for a color change, the puddle will drop out.

You'll have a hole rather than a puddle.

Always use a welding rod with a diameter closest to base-metal thickness. If you have a metal shear and you're welding sheet stock, filler rod can be made by shearing square strips from the sheet. This will automatically give you the precise filler-rod size and alloy.

For best results, the aluminum pieces must be absolutely clean. You should even wipe the welding rod with a clean, white cloth before welding.

Aluminum welding rod must be used with flux. Otherwise, you won't be able to weld aluminum successfully. Flux must be used to remove and inhibit the harmful oxides that form on aluminum. Mix aluminum welding-flux powder with water or alcohol. While welding, frequently dip the welding rod into the mixture to keep the rod coated with fresh flux. Brush on the flux every two or three minutes. The base metal must have fresh flux on it while welding.

Keep a large bucket of fresh, clean water nearby. After cooling, dip the parts in the water to rinse off the flux. Flux left on aluminum causes corrosion.

Oxyacetylene-Welding Aluminum—It's not absolutely necessary to use hydrogen to weld

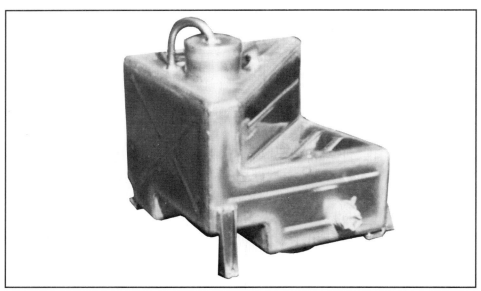

Ron Fournier gas-welded this oil tank for Ford GT LeMans race car. All *weldable* fittings are TIG welded instead of gas welded. Photo by Ron Fournier.

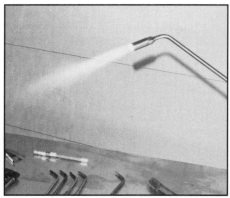

When rosebud tip is lit and adjusted, it should look like this.

ALUMINUM WELDING Q & A

A handy question-and-answer pamphlet for welding aluminum was prepared for the annual Experimental Aircraft Association (EAA) Fly-In at Oshkosh, Wisconsin. This pamphlet was based on the most commonly asked questions concerning aluminum welding by the 100,000 EAA members. With their permission, portions of the pamphlet follow:

Question: What aluminum alloys are weldable?

Answer: 1100, 3003, 3004, 5050, 5052, 6061 and 6063 are weldable. Specifically, 1100 is dead soft and not good for structures; 5052 is medium hard and good for fuel tanks; and 6061-0 is soft, but can be heat-treated after welding to make it very hard and strong. 3003, 3004, 5050 and 6063 are weldable, but seldom used. You can weld 3003, 3004, 5050 and 6063, but don't use these for your project. Stick with 5052 and 6061 for better results.

Question: What kind of rod should be used?

Answer: 1100 rod for 1100 material and 4043 rod for all other alloys.

Question: What flux should be used?

Answer: Antiborax #5 for cast aluminum and #8 for sheet aluminum.

Note: There are many other fine aluminum-welding fluxes.

Question: At what temperature does aluminum melt?

Answer: Pure aluminum melts at 1217F (658C)—less than half that of steel—but alloys melt at lower temperatures. Aluminum oxide, a corrosive film that forms on aluminum immediately after cleaning, melts at a much higher temperature than aluminum. This oxide must be *removed before welding and inhibited during welding.* Remove oxide with a stainless-steel wire brush, Scotchbrite abrasive pad or acid. Use flux before and during welding to prevent oxide formation.

Question: What equipment is needed to weld aluminum with oxyhydrogen?

Answer: You need a standard gas-welding torch, one oxygen regulator and cylinder, another *oxygen* regulator converted for use on a hydrogen cylinder and a cylinder of hydrogen. Converting the regulator involves changing the inlet and outlet fittings to left-hand fittings so you can attach the acetylene hose to the regulator and the regulator to the hydrogen tank. You should also use *cobalt-blue lenses* in your welding goggles. With the conventional green lenses, all you would see of the flame and weld puddle would be a large yellow spot. The blue lens filters out the yellow light blocking your view.

Question: What size welding tip should be used for welding aluminum with oxyhydrogen?

Answer: Use a tip three times larger than the one used for welding 4130 steel of the same thickness. For example, if you would use a #1 tip for welding 4130 steel, use a #4 tip for welding aluminum of the same thickness.

aluminum—just easier. For occasional aluminum welds, you can do an adequate job with an oxyacetylene-welding setup. As with oxyhydrogen, keep the tip clean. Always use a neutral flame, as opposed to an oxidizing flame, to prevent oxidizing the aluminum. Also, use *cobalt-blue* welding lenses to filter out yellow glare.

Line Pressures—Use as low a line pressure as you can, such as 2—3 psi with a Victor #2 or equivalent tip. This helps to avoid overheating and blowing the aluminum away. Aluminum dissipates heat much quicker than steel, requiring the use of a large tip to maintain heat. The difficulty arises out of using 6300F (3482C) to keep a molten puddle just above 1217F (658C).

HEATING & FORMING

When you heat steel to dark blood red (1050F or 566C), it bends easier than when cold. Heat steel to bright cherry red (1375F or 746C) and you won't believe how easily it bends! Steel molecules are more *plastic,* or pliable, the hotter they get.

You can use heat to assist in bending large, thick pieces of steel with a simple bench vise and a small pry bar. I modify new trailer hitches to fit old cars simply by heating and bending them. I learned that trick when I was about seven years old, helping my uncle make new plow tips for his farm tractor. Red-hot steel bends like warm taffy candy.

Before lighting gas torch with a rosebud tip, adjust oxygen pressure at regulator relatively high. I've adjusted it to 27 psi.

Also, increase acetylene pressure when using rosebud tip. Set regulator at 6-10 psi. Never exceed 15 psi. Note dark line—it's red—starting at 15 psi on low-pressure gage.

When separating stuck steel parts, such as this U-joint and stub axle, mark part to be heated using a 350F temperature-indicating crayon.

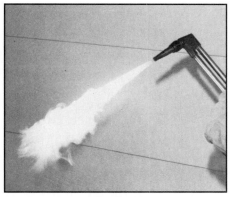

Just-lighted, acetylene-rich cutting torch looks like this. Acetylene valve is open about one turn to minimize sooty flame.

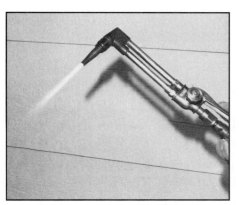

Opened cutting-tip oxygen valve neutralizes flame. Remember: Oxygen valve on torch body is fully opened.

Practice bending hot steel and pretty soon you'll be doing things you never thought possible. Just use the appropriate-size welding tip, cutting tip or *rosebud* tip to apply the amount of heat needed. I try to use a tip that heats the metal to cherry red in less than one minute. If it takes longer, the tip is too small. This wastes both time and gas.

Rosebud Tip—A rosebud tip is so called because its flame configuration looks like a rosebud. Strictly meant for heating, this tip is useful for heating and forming metal. A rosebud tip is no good for welding, but it does generate gobs of heat. It also uses a lot of oxygen and acetylene, so make sure your tanks aren't low before

starting a project.

The first time I lit a rosebud tip, it produced a *pop* that sounded like a shotgun! The next six times I did it, I got the same loud noise! It occurred to me that 4-psi oxygen and acetylene pressure were insufficient to operate that big tip! I then increased oxygen pressure to 25 psi and acetylene pressure to 10 psi; the torch lit with only a soft pop. It takes a lot of pressure to operate a rosebud tip!

Rosebud tips are good for freeing stuck parts. However, because of their high-heat output, use a temperature-indicating crayon or paint to check the temperature so you don't overheat the part.

Cutting-Torch Tip—If necessary, you can use a cutting-torch tip for heating. It puts out more heat than a welding tip, but less than a

rosebud tip. Adjust the torch for the maximum neutral flame *without* the cutting lever on. Be sure to not hit the cutting lever while you're heating a part, or you'll oxidize or cut it!

GAS CUTTING—OXIDIZING

Now that you've "mastered" the art of gas welding, the next item on the list is learning to flame-cut with acetylene—sometimes referred to as a *blue wrench* or *gas wrench*. These terms came from mechanics who used cutting torches for loosening or removing stubborn bolts, nuts or other seized parts, usually due to rust. Our main purpose is to use the acetylene cutting torch for fabricating steel parts.

The primary chemical reaction in flame-cutting steel is oxidation. Because cutting is really oxidizing,

Preheat edge of metal and make puddle before depressing oxygen lever.

When puddle forms, depress oxygen cutting lever and move torch along cut-line. On 3/8-in.-thick steel, move about 1 in. every three seconds.

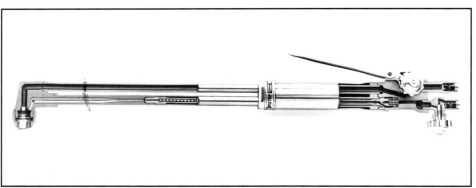

Cutaway of cutting torch: This style cutting torch does not use separate torch body. Photo courtesy of Victor Equipment Co.

you cannot cut metals that do not oxidize (rust) easily, such as aluminum and stainless steel. If you cut 4130 steel, you over-oxidize and damage the heat-affected area, causing cracks later on.

So primarily, you can flame-cut mild steel and cold-rolled steel. These steels make up a large part of things we use, such as automobiles, trailers, farm equipment and more. Flame-cutting is useful for cutting elaborate shapes not suitable for cutting with a bandsaw. Remember that flame-cutting leaves rough edges with slag that must be final-trimmed later. This is useful if you're doing arts and crafts, but a problem when doing precision fitting.

USING A CUTTING TORCH

After learning to gas weld, you will discover that a cutting torch is relatively easy to operate. Simply light the torch, adjust the *flame* to neutral, make a little puddle, push the oxygen lever and you're cutting steel! Here are the steps for cutting 1/4-in.-thick steel plate:

Select a piece of 1/4-in. scrap steel plate. Also find a piece of angle steel similar to the one shown. Use it to guide your torch-hand and help you cut a straight or smooth curved line. Secure it with C-clamps or locking pliers, if necessary. Mark the cut-line with soapstone.

Position the plate so the cut-line hangs over the edge of the welding table. Or, lay two short sections of angle iron face down on the bench and lay the plate on top. This will prevent a nice cut being made across your workbench top.

Before you light the torch, check your clothing. You should be wearing cuffless pants, high-top shoes, a long-sleeve shirt, gloves and welding goggles. If all is OK, proceed.

Light Cutting Torch—Adjust the regulators for about 1—2-psi acetylene and 10—15-psi oxygen. Oxygen pressure is much higher than acetylene pressure because the oxygen does most of the work.

Next, preadjust the cutting tip. Shut off the oxygen valve on the cutting tip and fully open the oxygen valve on the torch handle. This is important. Open the acetylene valve about one turn and light the torch using a flint or electric striker. Add oxygen by opening the oxygen at the cutting tip until you have a neutral flame.

Start Cutting—If you're right-handed, support the torch with your left hand while resting it on the plate. This will allow you to guide the torch along the cut-line for a more-even cut. Grip the torch with your right hand and have your thumb ready to press the oxygen—cutting—lever. Start your cut by heating the edge of the steel plate at the right end of the cut-line. Reverse everything if you're left-handed.

Note that when cutting with acetylene, material is removed. The resulting void is the *kerf.*

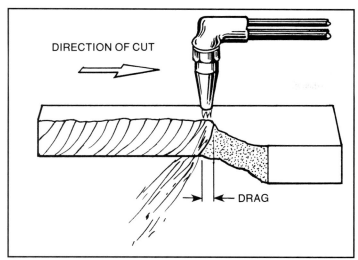

If speed is too high or oxygen flow not enough, *drag* occurs when cutting thick material. Reduce or eliminate drag by reducing speed or increasing oxygen flow—pressure.

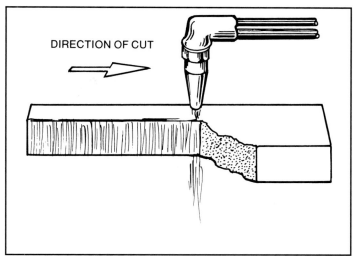

Drop cut occurs when cutting speed and oxygen flow are correctly matched so there is no drag. Flame exits kerf immediately below where it enters.

Consequently, you must cut on the *outside* of the cut-line—scrap side. Otherwise, the part you're making will be short by the kerf width.

When a puddle develops, press on the oxygen lever. The flame should immediately begin to blow away—oxidize—the metal. If it doesn't, you didn't have a good puddle. Continue heating the metal and try again. When the sparks fly as the cut begins, carefully guide the torch along the cut-line. With a 1/4-in. steel plate, you move along at about 1 in. every three seconds, following the cut-line. Move too fast and you'll get a shower of sparks back in your face because the metal wasn't heated enough. Move too slow and you'll overheat the metal, resulting in excessive slag.

Be careful when you reach the end of your cut. A chunk of hot metal may fall to the floor if you've made a clean cut. But, chances are the slag will hold the pieces together. A light tap with a hammer should break them apart.

TIPS FOR BETTER FLAME CUTTING

- Excess slag at bottom of cut indicates the preheat flame is too hot. Correct by reducing acetylene pressure or using a smaller tip.

It took about 40 seconds to make cut. Note scrap piece falling to floor. Watch the toes!

- Metal doesn't have to be super clean for cutting.
- Most beginners force the cut by moving too fast. Slow down. Refer to the chart for ideal cutting speeds for different metal thicknesses.
- Clean cutting tip periodically. The cutting process tends to splatter molten metal back on the cutting tip, reducing cutting efficiency.

Cut has too much slag on bottom of kerf (arrow), indicating excess acetylene pressure. I used 4-psi acetylene; 2 psi would've been better.

FLAME-CUTTING AIDS

There are a few items you can use that'll make a cutting project easier. They range from a piece of angle iron used as a guide to cut a straight line to a sophisticated flame-cutting machine. Chances are you won't need the cutting machine unless you'll be duplicating

CUTTING STEEL WITH OXYGEN ONLY, NO FUEL!

Here's a trick I show new welding students to illustrate that cutting with oxyacetylene is primarily an oxidizing process. Try it after you learn to use the cutting torch.

Start your cut in a piece of steel plate about 1/4-in. thick. After establishing the cutting speed and travel, shut off the acetylene torch valve with your left hand—right hand if you're left-handed—while you continue to cut. You'll still be able to cut the steel with oxygen only! Of course, if you don't move steadily, you'll lose the molten puddle and the oxidizing or cutting process will stop. You'll then have to relight the torch to resume the cut.

SETUPS FOR CUTTING STEEL										
Material Thickness (in.)	1/8	1/4	1/2	3/4	1	1-1/2	2	4	5	6
Suggested Tip Number	00	0	1	1	2	2	2	3	3	4
Oxygen Pressure (psi)	5 to 10	10 to 15	10 to 20	15 to 25	20 to 30	25 to 35	30 to 40	35 to 45	40 to 45	45 to 50
Acetylene Pressure (psi)	1 to 2	1 to 2	2 to 3	2 to 3	2 to 4	2 to 4	2 to 4	3 to 5	3 to 5	4 to 6
Cutting Speed Per Minute (in.)	20 to 23	18 to 20	14 to 16	12 to 16	10 to 14	8 to 12	6 to 10	5 to 8	4 to 6	3 to 5
Oxygen Used Per Hour (cu ft)	50	80	120	140	150	170	220	350	420	500
Acetylene Used Per Hour (cu ft)	8	10	12	15	16	17	19	25	29	30

Refer to chart when setting up to cut.

Welder is using angle steel as guide to make straight cut. Slag and sparks are contained in metal bin below cutting table.

Bevel-cut is being made on 3/4-in.-thick steel plate. Motor-driven, track-mounted cutting torch gives a smooth cut because travel speed is uniform. Straight and curved tracks are available. Torch head is fully adjustable.

several pieces from heavy steel plate.

Marking Tools—I already discussed soapstone, center punches and scribes for laying out and marking cut-lines, page 40. To assist with marking, you should make patterns. A pattern duplicates the shape you want to cut out. It can also be used for checking the fit of the final part without actually having the part.

Once you've *developed* the pattern, it is laid on the material to be cut out for the final part and traced around with soapstone or whatever marker you desire.

Pattern material is inexpensive. All you need is plenty of thin, flexible sheets of cardboard. Such material is available at office-supply stores. Corrugated cardboard from boxes is cumbersome but OK to use in a pinch. To mark patterns, you need soft-lead pencils and a pencil compass. Additional drawing equipment such as straightedges, 30°/60° and 45° triangles and felt-tip markers should also be on your list of pattern-making items. Finally, you'll need a pair of heavy-duty scissors for cutting-out patterns.

Cutting Table—If you'll be doing a lot of flame cutting, it would be helpful to have a cutting table. See the drawing, page 139. A cutting torch can't distinguish a metal-top welding table from a workpiece. Consequently, you shouldn't flame-cut anything that's laying directly on top of the table. Otherwise, you'll end up making a cut through the table top as well.

To avoid this, either raise the workpiece off the table by supporting it with scrap angle iron, hang it over the edge of the table or set it somewhere else. Whatever you do, don't support it with anything you don't want to be cut or damaged. Ideally, it is best to support the work with a cutting table.

A flame-cutting surface doesn't have to be exotic. It can be as simple as several sections of angle iron, positioned corners up, bridging two metal saw-type horses. Or, it can be an honest-to-goodness table. The table would consist of an angle-iron frame with four legs. Instead of using a solid steel top, string several sections of angle iron loosely between the frame. This will allow you to adjust the angle iron to support the work as

desired. You can also replace them easily when they are cut.

Cutting Machines—Cutting machines used in the welding industry can flame-cut parts to exacting dimensions, duplicate parts or cut out many parts at one time. Some of this equipment is very expensive, such as an electric-powered machine that runs on its own track. More expensive machines have multiple cutting heads. While a tracer follows a single pattern, the multiple cutting torches cut out several duplicate parts simultaneously.

Chances are you won't need such equipment. For the hobbyist or small welding shop, inexpensive

Cutting machine is designed to fit on 55-gal drum. Tracer follows pattern, above, while torch cuts workpiece that's on grate. Photo courtesy Williams Low-Buck Tools.

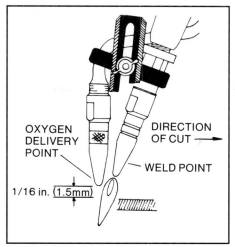

OXYGEN DELIVERY POINT

DIRECTION OF CUT →

WELD POINT

1/16 in. (1.5mm)

Preheat flame is separate from oxygen-delivery tip on Dillon MK III cutting torch. Consequently, cut *must* be made with oxygen tip trailing preheat tip as shown. Oxygen tip is adjusted to other side of preheat tip for cutting sheet metal. Drawing courtesy Shannon Marketing, Inc.

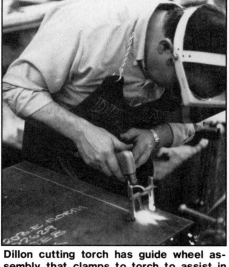

Dillon cutting torch has guide wheel assembly that clamps to torch to assist in making precise cuts. Torch must be turned to follow cut-line. Photo courtesy of K. Woods, Inc.

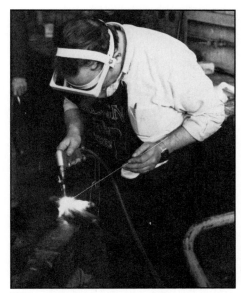

With cutting tip removed, Dillon torch can be used for welding steel, stainless steel, cast iron, brass, copper and aluminum. It can also be used for brazing, soldering and cutting. Photo courtesy of K. Woods, Inc.

flame-cutting machines for cutting out small parts are available, such as the one shown above from Williams' Low-Buck Tools. Manually operated with a tracer, this machine is designed to install on a 55-gallon drum. The drum contains all the sparks and scrap that would otherwise fall to the floor.

DILLON MK III TORCH

They say there is nothing new under the sun, but there are different ways of applying a principle. The Dillon MK III is a new-design gas-welding torch made in Australia. You hold it like a pistol, unlike the more conventional welding-torch design.

Cutting—Traditional gas torches use an oxyacetylene cutting tip with six small flames in a small circle to preheat the steel, and a large orifice in the middle of the six smaller holes. The larger orifice is where the stream of oxygen comes out when you depress the lever on the torch body for cutting. This means that no matter which direction you travel with the cutting torch, you follow the nice, clean-cut *kerf,* page 60, with

at least one or two more oxyacetylene flames that tend to melt the metal back together.

The Dillon cutting torch attachment is different. It uses a single oxyacetylene flame to preheat the steel and a single oxygen-only stream of higher-pressure gas to make the cut. No acetylene flame follows, so the cut stays clean. But you don't get something for nothing. The Dillon cutting attachment works only in one direction. If you want to cut a circle, you have to move the cutting attachment in a circle to keep the preheat oxyacetylene flame and oxygen in a leading-trailing relationship.

Cutting thin sheet metal, such as car fenders, is really where this torch excels. For sheet-metal cutting, use a different attachment that places the oxygen-only tip at the rear of the flame so you make the cut going away from you. This allows very thin sheet metal to be cut quickly and with almost no heat-affected zone adjacent to the kerf. This feature is ideal for the new high-strength steels that tend to crack in the heat-affected zone.

Welding—The Dillon torch also

works well as a regular gas torch. The factory brochure recommends torch settings of 4-psi oxygen and 4-psi acetylene, but I learned to weld with much lower pressures than that. In the oxyacetylene pressure chart on page 46, I recommend 4 psi for #4 tips, but only 1 psi for a #1 tip. Anyone who uses more than 4 or 5 psi to do gas welding will surely oxidize the metal and burn up the weld, no matter what brand torch is used.

TEMPERING & HEAT-TREATING

Don't throw away that old screwdriver just because the blade is dull or twisted. With your gas-welding torch, you can reshape the blade, heat-treat it, and temper it to like-new. While you're at it, gather up all of your dull steel chisels and recondition them, too.

You'll need a friend to hold the torch. You can then hammer the red-hot screwdriver or chisel tips and *quench*—submerge-cool—them in oil. Required equipment includes a heavy-duty bench vise with an anvil pad, an anvil, or a heavy piece of flat steel such as an old body-and-fender dolly. You'll use this as backup for hammering the screwdriver and chisel blades. Other necessities include a medium-size ball-peen hammer, a bench grinder for shaping the blades, and a quart of 10W motor oil. Of course, you'll also use your oxyacetylene torch for heating and tempering.

Reshaping the Blades—Using a #3 or equivalent tip on your torch, heat the flat portion of the screwdriver blade, chisel or whatever from the tip to about 1 in. up the shank to bright cherry red. As soon as it's bright cherry red, hand your helper the torch and hold the blade on the "anvil" while you lightly tap it slightly oversize with a ball-peen hammer. This should take no longer than 5—10 seconds and 10 or so taps with the hammer.

Now, quickly quench the heated tip in oil so heat going up the shank will not melt or burn the handle. Next, wipe clean the blade and shank and reheat the tip until it almost turns yellow, then immediately quench the tip and shank in oil. The oil will bubble and boil. Swish the screwdriver or chisel in the oil until it is cool, then remove it and wipe the tip clean. At this point, the blade is rough heat-treated, but obviously not ready to use.

The next step is to dress the flat surfaces to proper shape with a grinding wheel. Make sure the blade tip is square to the grinder. For a screwdriver, grind until the tip fits the screw slots it's intended for. Be sure to grind the sides, tip and face of the blade so no dull metal remains.

Be careful when grinding. If you overheat the metal, you'll have to retemper it. So, grind a little, then cool the tip by dipping it in water. Continue grinding and cooling until the tip is fully dressed.

The last step is the actual tempering, or *drawing*, to remove brittleness—not hardness. In this step, carefully bathe the newly shaped blade tip in an oxyacetylene flame so the shiny metal changes to the color given below. Only the extreme tip of the flame must touch the metal. The color the tip is heated to indicates a specific tempering temperature required for different tools. It is easy, try it—but don't get the metal too hot.

Tempering colors are:

Item	Temperature F (C)	Color of Steel
Screwdriver	600 (316)	pale blue
Ax, chisel	550 (288)	violet
Center punch	525 (274)	purple
Wood chisel	500 (260)	yellow-brown
Lathe tool	375 (190)	light yellow

With any gas-welding torch, you should maintain a neutral flame with equal oxygen and acetylene pressures, and a flame that is neutral or just *off secondary feather.* As a general rule, use as low a pressure as your welding torch will maintain. Any good gas-welding torch will weld cast iron, aluminum, stainless steel, copper, brass and mild steel, if you observe the guidelines in this book.

Remember, before you buy any welding equipment, try it and compare it.

7

Gas Brazing & Soldering

Taking a victory lap at Phoenix International Raceway in my small-displacement sports-racing special. Car uses brazed mild-steel-tube frame, page 66.

Brazing and soldering are metal joining methods that do not rely on melting the base metal to join two or more pieces. Instead of fusing the filler and base metals, they depend on surface adhesion of the solder or braze filler. This is made possible by *capillary action*—surface of the molten filler is attracted to fixed molecules nearby. When the filler metal cools, it bonds to the base-metal surface.

Brazing and soldering are more akin to adhesive bonding than welding. A major advantage of these processes over welding is that brazing and soldering are done at lower temperatures. Soldering is done below 800F (427C); brazing is done below the melting point of the base metal,

usually below 1500F (816C). Therefore, warpage and temperature-induced stress in the base metal are lower. For example, to fusion-weld a bicycle frame, you must heat the metal to the melting point of steel—more than 2700F (1482C). To braze that same frame, 1000F (538C) is all that's necessary.

Another big advantage of brazing and soldering is that field repairs are simple. All you need is a torch. To arc- or TIG-weld in the field, such as repairing farm equipment, you'd have to perform the operation near an electrical outlet or a portable welder. That's not always easy to do.

In case you were wondering, brazing does not mean *brass* any more than soldering means lead

or silver. For instance, there is brass, aluminum and silver brazing. Brazing refers to the *temperature,* not the metal.

Most people think that brazing isn't as strong as arc welding because they've seen or heard of brazed joints breaking. To disprove this, compare the tensile strengths of welding and brazing rods. E-7018 is considered to be the best arc-welding rod and it has a 70,000-psi tensile strength. Now, look at the tensile strength of several brazing rods on page 117—80,000 psi or more!

To demonstrate the tensile strength of a brazed joint, I often ask my welding students to guess how much their cars weigh. The answers usually range from 2500 to 5000 lb. You probably know

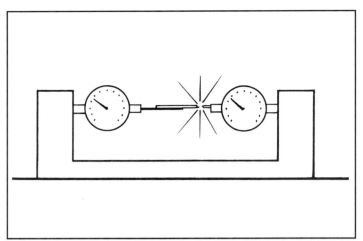

Laboratory pull-test machine can test spot welds, braze joints and solder joints. Gage indicates tensile strength of part or weld being tested.

Race-car frame was brazed for two reasons: Low-temperature joining method reduces metal stress and warpage; and repairs at track can be made more easily.

what's coming next. I then take two 1 X 5-in. long, 0.060-in.-thick mild-steel strips, overlap the ends 1 in. and braze them together.

Next, I put the brazed strips in the welding-shop pull-test machine and let the class watch as the machine pulls over 3000 lb before the *base metal* stretches and breaks! The brazed joint never breaks. Instead, the metal at one end of the joint fails. By this time, the whole class is saying, "That little 1 sq in. of brazing could lift that entire 3000-lb car!" More importantly, the brazed joint is stronger than the metal itself!

Seam or Joint Design—The design of a seam or joint is very important in deciding whether to use solder or braze. For instance, avoid brazing or soldering butt joints. The main reason for this is the lack of sufficient *wetted* base-metal surface area for filler to adhere to. Consequently, the base metal pulls away from the braze or solder under *tension*—pulling—or bending loads. Lap joints are another matter. The wetted surface area can be increased by simply increasing the overlap of the two pieces. And the joint is in *shear*—the best type of joint for brazing or soldering. See the illustrations of types of joints and loads, page 69.

Later in this chapter I describe which joints are best for *brass*

Gas-fusion welding was used to join pipes of race-car exhaust header, but brazing joined flanges to pipe. Holes are too easily burned in thin-wall pipes when fusion-welding them to thick mounting flanges.

brazing, silver brazing, silver soldering and *lead soldering*.

BRAZING

Because brazing requires less heat, and therefore results in less warpage than fusion-welding, brazing is used extensively in auto-body repair. Less heat also allows you to braze near rubber parts because of the reduced chance of burning the rubber.

The lower heat and induced stresses also mean that you don't have to be as careful about avoiding air currents when brazing as you do when fusion-welding sensitive metals such as 4130 steel tubing. And materials of different thicknesses can be joined easily. You don't burn up the thin part trying to heat the thick part.

Brazed joints look good and usually require only flux removal to maximize their final appearance. Brazing, by its nature, will flow into a smooth fillet, giving the joint a finished look without filing or machining

Remember: *Never fusion-weld a joint that was previously brazed.* If you have a project that requires

Disc-brake adapter flange was brazed rather than welded to front-wheel hub because metals are dissimilar. Both parts were preheated to 350F (177C) before brazing.

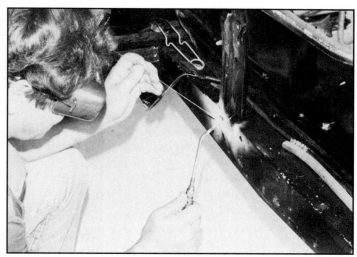

To braze sheet-metal brace to Corvair rear frame, I'm holding torch about 6 in. from metal for a softer and cooler braze joint. After braze is completed, smoke and flux residue can be brushed or washed off with water.

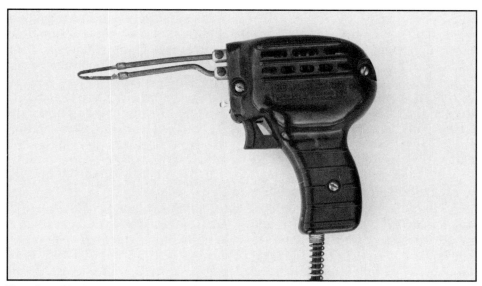

Electric soldering gun is not suitable for most soldering jobs, only very light-duty tasks such as joining wires. Instead, use your gas torch with a soft flame. Photo by Tom Monroe.

both fusion welding and brazing, fusion-weld first, then do the braze joint. Never braze first. Otherwise, you'll boil away the braze filler with the higher temperature of the fusion weld.

You can maximize the strength of a brazed joint by giving the joint more surface area. Depending on base-metal thickness, roughen it by grinding, sanding, sandblasting or coarse filing. All those little surface scratches increase surface area and provide a *tooth* the filler can cling to.

Identify the base metal and use the correct filler metal. For instance, if you tried to braze shiny stainless steel with shiny aluminum filler rod, the metals just would not mix. You'd end up with a hot mess that will fall apart.

Always heat the base metal sufficiently, especially cast iron. But don't overheat small areas. This may cause the braze filler to fume and boil.

Metals That Can Be Brazed—A wide variety of metals can be joined by brazing. These include stainless steel, cast iron, brass, copper, bronze, aluminum (with aluminum brazing rod), mild steel, cold-rolled steel, chrome-plated metal, cast metal, galvanized steel and other zinc-coated steels.

Dissimilar metals can be joined, such as copper to steel, or copper to brass. This usually is not possible in fusion welding.

Brazing Cautions—You must *never breathe brazing vapors*. Use ventilation as necessary. When galvanized or zinc-coated metals are heated with a welding torch, fumes are given off that are extremely dangerous if inhaled. These fumes appear white, similar to cigarette smoke. It's safest to avoid welding, cutting, brazing or soldering galvanized or zinc-coated metals. Instead, let an experienced weldor do it for you. But, if you feel you must weld galvanized pipe or sheet metal, follow these precautions:

● Weld galvanized metal only outdoors in open air so fumes will not concentrate. Or . . .

● Use a commercial air extractor to suck the fumes into a filter and away from humans and animals.

● Wear a high-quality breathing respirator while welding galvanized metal. See the suppliers list in the back of the book for addresses.

GAS-BRAZING PROCEDURE

When practicing brazing, you'll be doing similar, but fewer, operations than you did when practicing gas-welding steel. Brazing involves three operations instead of five. You won't puddle brass or do butt welds. Instead, you'll concentrate on lap welding.

You'll need several 2 X 5-in. pieces of 0.030-in.-thick mild-steel scrap. Your three projects are:

- Running a bead with brazing rod.
- Lap-brazing two pieces.
- T-brazing two pieces.

Required Equipment:

- Oxyacetylene-welding outfit.
- 1/16-in. brazing rod (36-in. long, cut in half). For the required filler material, see page 117.
- Powdered brazing flux or flux-coated rod.
- Bucket of water for removing flux from braze bead.

Common Mistake—Many inexperienced braze weldors overheat the base metal. After learning to fusion-weld mild steel at 2700F (1482C), they must learn to braze at 1050—1075F (566—580C). Remember that all metals start to vaporize at their boiling point. Brass brazing rod will boil, vaporize and generally ruin your project if heated to the melting point of steel!

For best results when brazing mild steel, heat the base metal to about blood red to *dark* cherry red—no hotter! To do this with a 6300F (3482C) gas-welding torch, hold it farther away from the metal than if you were fusion-welding. You may think that a smaller torch tip will give less heat—true. But, for brazing, you should use a large *soft* flame rather than a small *hot* or *harsh* flame.

To get a *soft* or *quiet* flame, use a medium tip with very low gas pressures. You'll hear the difference between the *soft* flame compared to a neutral flame for welding steel. You'll like the way the soft flame sounds.

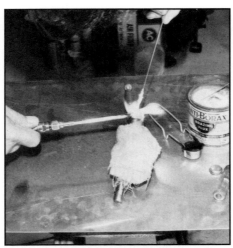

Note can of brazing flux nearby and flux-coated rod. To add flux, heat end of rod, then dip into flux. Shift tube was repaired by brazing shut worn shift-pin holes and re-drilling to size. Wet cloth wrapped around rubber bushing prevents heat damage.

Flux the Rod—Open the can of brazing flux. Using a 0.040-in. tip, set both oxygen and acetylene regulators to 2—3 psi. Before you light your torch, double-check that you have on the proper apparel: long-sleeve shirt, long pants, gloves and welding goggles.

Light the torch and adjust for a soft flame; not loud and hot. *Brush,* or bathe, the working end of the brass rod with the flame to get it warm—not molten. Quickly dip the rod into the powdered flux. When you pull out the rod, flux should have adhered to about 2 in. of the hot end of the rod, completely covering it. *Use plenty of flux.* Don't worry about using too much. A can of flux goes a long way. One can lasts me years, even when I do a lot of brazing.

Precoated Rods—You can buy brazing rod precoated with flux, eliminating the inconvenience of repeatedly dipping the rod into the flux. Sometimes, even an old pro like me buys some.

The problem with flux-coated rods is that they must be protected from moisture and rough handling. The flux coating can break and flake off. If this happens, there won't be enough

flux on the rod to do a good braze job. I buy just enough rod to last 30 days or less, so the rods are always fresh.

Run a Braze Bead—Set your piece of scrap metal on firebricks. If you haven't already done so, light the torch, adjust it to a soft flame and hold the tip 2—3 in. from the workpiece metal until a 1-in.-or-so round blood-red spot develops. Now, just touch the hot spot with the brazing rod. The filler should melt and flow onto the steel. Continue this heating and touching process until the rod needs more flux. Dip the rod back into the flux and keep going.

BRAZING JOINTS

Lap Joint—Now you're going to braze a lap joint and see capillary action at work. Make sure your two pieces of scrap steel are clean and rust-free. They should be flat so the edges will not bow up and look bad afterward. See the accompanying drawing for how to fit the two pieces. Remember to support the workpiece off the table so it doesn't soak up the heat.

Light the torch as before and coat the rod with flux. Play the torch along one end of the seam to heat both pieces to dull cherry red. When you're *sure* that *both* pieces are the same color and temperature, touch the edge of the seam with the flux-coated brazing rod. Watch the molten filler flow into the seam! That's caused by capillary action. Continue along the seam until you have it filled end-to-end with filler.

Shut off the torch and let the metal cool for about 3—4 minutes. Pick up the metal with your pliers and dip it into a bucket of water to cool and soften the flux. The flux turns into a glass-like substance after cooling from its molten state. Although you can chip it off, it's easier to let water soften it. Then, it's an easy job to remove the flux with a wire brush.

If you succeeded in getting the filler to flow completely into the seam, the joint should be capable

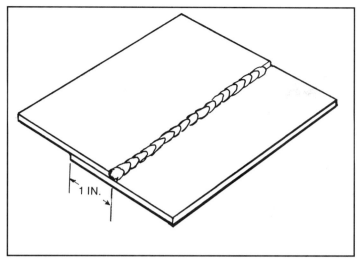

Overlap braze lap joint about 1 in. Heat joint evenly until both pieces are blood red. Apply brazing rod and watch it flow into joint.

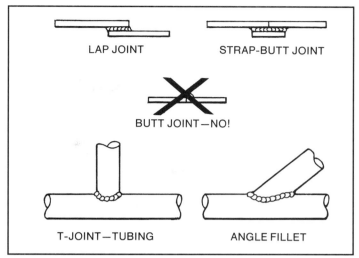

LAP JOINT

STRAP-BUTT JOINT

BUTT JOINT—NO!

T-JOINT—TUBING

ANGLE FILLET

Best and worst joints for brazing, silver brazing or soldering: *Never* braze butt joints.

of lifting a car!

T-Joint—After you've mastered brazing the lap joint, practice brazing a T-joint. Support the pieces so heat isn't absorbed into the work table. Hold the T in place with a "mechanical finger"—page 52. *Tack-braze* it in place. A tack should go at each end of the T.

As you start brazing the seam, remember to control the temperature of the pieces. Remember gas-welding a T-joint? You cannot heat just the bottom metal piece and expect the brass to flow onto the vertical piece. *Manipulate* the torch so both pieces are heated equally—dark-cherry red. Remember, if you overheat the steel, the brazing rod will fume or boil! After you've finished the seam, shut off the torch.

The seam must be filled with filler to be strong. Adhered surface area must be maximized. The above drawings show the best seams for brass and silver brazing.

Brazing Aluminum—Brazing aluminum is similar to that for brazing steel or other materials. It also has the same problem as gas-welding aluminum: The base-metal color doesn't change as it's heated.

The clue to judging correct temperature for brazing aluminum is to watch the flux that's on the base metal. When it starts to melt

End cap on chrome-plated intake manifold was fusion-welded, then mounting flange was brazed. It couldn't have been done in reverse order because braze metal would have boiled while fusion-welding end cap.

and flow, the base metal is ready to be brazed.

Use a *slightly* carburizing flame to reduce aluminum oxidation.

Not only must aluminum-brazing rod have flux on it, you must also apply flux to both sides of the weld joint. I like to use a small metal-handled *acid brush* to apply the liquid flux, but you could even paint it on with a thick-bristle paint brush. Don't use a brush with plastic bristles—the bristles will melt. Acid brushes are

relatively inexpensive—so inexpensive that you could throw them away after one use. For this reason, I keep a dozen or so brushes around at all times.

Brazing Copper, Cast Iron & Other Metals—By the time you're ready to braze other metals, you should be able to determine when the base-metal temperature is right for brazing. Copper turns red, stainless steel blue, and cast iron yellow when they've reached the right

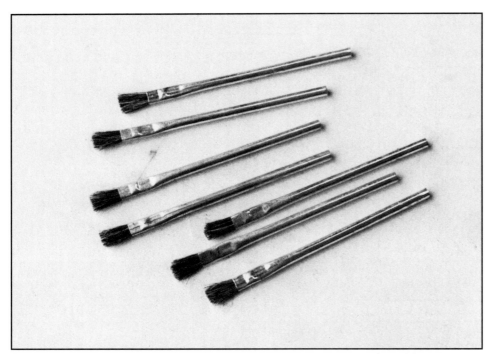

Flux is applied to joint with acid brush. Brushes are inexpensive and used just once, so you'll need many. Photo by Tom Monroe.

Silver-soldering baffles to Corvair engine oil-pickup screen: Higher-temperature brass brazing might have burned oil-pickup screen inside.

temperature. But, as with brazing aluminum, the best way to judge when the proper temperature has been reached is to watch the flux. When it melts, the base metal is ready to accept the filler metal.

SILVER BRAZING

You can silver-braze just about any metal that can be brass-brazed. Silver wets the metal better than brass. It also sticks to some metals where brass will not, such as *carbide tool steel.* This very hard steel is used for tipping saw blades and other cutting tools. Silver braze also gives a superior appearance to some projects such as costume jewelry.

Because silver is one of the metals in silver-brazing rod, it costs much more than brass filler. A silver-bearing alloy of low-tensile strength—about 20,000 psi—can be used to join dissimilar metals such as aluminum, steel, copper, stainless steel and monel. This particular low tensile-strength silver alloy melts at low temperature—about 500F (260C).

Silver-Brazing Procedure—The first thing to do is thoroughly clean the joint surfaces. The joint clearance should be 0.002—0.006 in. If you can't judge this spacing by eye, use a feeler gage to set the spacing at exactly 0.004 in. After setting up parts with a feeler gage a few times, you should be able to judge the 0.002—0.006-in. gap by eye with ease.

Paint the joint area with flux thinned with water or alcohol. Coat both sides. Use a slightly carburizing flame and heat a broad area, keeping the torch in motion. When the flux turns clear and starts to run, add enough silver alloy to completely fill the joint. When finished, shut off the torch and allow the joint to cool for 3—4 minutes. Remove flux with hot water.

SOLDERING

Soldering is another welding operation done below 800F (427C). Lead soldering and silver soldering are similar processes. The major difference is that lead soldering is more akin to brazing because of the lower temperatures used. Brazing, lead soldering and silver soldering all require the use of heat, flux and capillary action.

Metals that are easily soldered are platinum, gold, copper, silver, cadmium plate and tin. Less easy to solder are nickel plate, brass and bronze. Metals that are more difficult to solder because they don't wet easily are mild steel, galvanized plate, and aluminum alloys 1100, 3003, 5005, 6061 and 7072.

Electric Soldering—Most people are familiar with soldering copper wire with an electric soldering iron or gun. I concentrate on flame soldering. The reason for this is that electric soldering generally is restricted to copper wiring and similar parts that have a small amount of mass. Although it's a low-temperature welding process, a conventional electric soldering iron, page 67, simply won't heat much more mass than a thumbtack to a temperature sufficient for soldering! The temperatures involved are the same, but the *quantity* of heat is different.

Soldering Procedure—Although welding goggles are not required to solder because intense light is not generated, you should wear safety glasses to prevent eye injury in case the solder pops or splatters. Once the goggles are on and the torch is in hand, follow these steps to flame solder:

● Clean the base metal with Scotchbrite abrasive pad, steel wool or emery paper. Remove

Before soldering copper water pipe, Breven cleans joint with Scotchbrite abrasive pad. Failure to clean pipe would prevent solder from adhering.

Flux should be brushed wherever solder should stick.

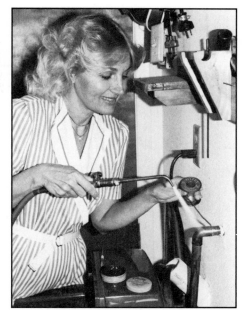

With a very soft flame, Breven lightly bathes copper tubing with tip of flame. When flux melts and starts to flow, solder is touched to joint, melting it. Capillary action pulls it inside joint. *Safety glasses should have been worn* for solder repair in case solder popped or splattered. Acid-core solder is OK for soldering water pipes, but use rosin-core solder for electrical work.

all oil, grease, paint and *anything* not part of the base metal. Otherwise, the solder will not adhere to the metal.

- Apply flux to the base metal. Choose the correct flux as suggested in Chapter 12. Use an acid brush to apply it to all surfaces you intend to solder.
- Heat the base metal to soldering temperature by playing a soft flame over the base metal until the flux melts and starts to run. Then touch the metal with

solder. If heated correctly, it will flow into the joint by capillary action.

- Apply solder until the joint is filled. Apply heat as needed.
- If necessary, remove flux from the joint. Use a wire brush or water for this.

For free information on brazing and silver soldering, write:

Chemetron Corporation
111 East Wacker Drive
Chicago, IL 60601

Ask for their booklet, "Construction Manual and Catalog."

Thermacote-Welco Company
32311 Stephenson Hwy.
Madison Heights, MI 48071
Ask for their booklet, "Welco Alloys Technical Guide."

Handy & Harmon
850 Third Avenue
New York, NY 10022
Ask for their booklet, "The Brazing Book."

Arc Welding

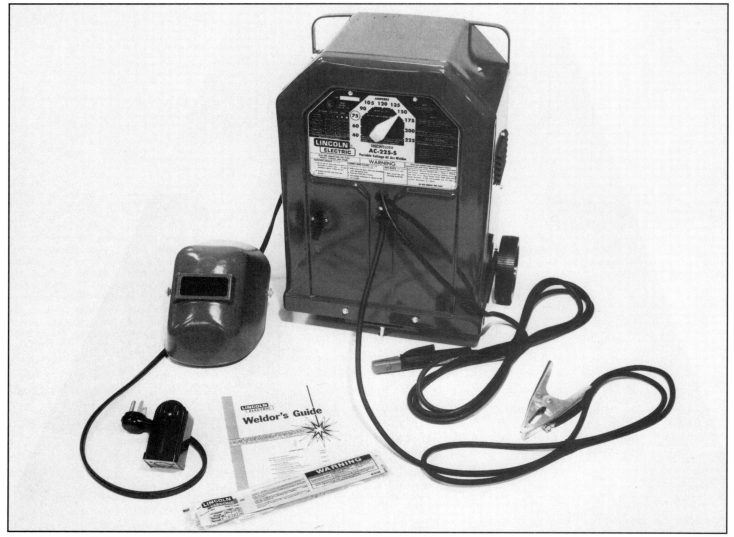

8

Buzz-box is suitable for small welding shop, farm or home workshop. This is the ever popular Lincoln 225-amp a-c arc welder. Photo courtesy Lincoln Electric Company.

Before reading this chapter, you should read about how to gas-weld, page 45. As mentioned throughout this book, gas welding should be mastered or, at least understood, before you attempt other welding techniques.

This section describes how to buy and set up the arc welder you need and can afford. I say "afford" because arc-welder prices range from $99 to $50,000. You probably don't want or need the cheapest or most expensive welder.

Just as with oxyacetylene equipment, you must determine your welding needs, then choose the arc welder that meets those needs. Remember, every welding project in Chapter 15 can be made with arc welders costing less than $200.

For most of you, a large-capacity arc welder is unnecessary. Even commercial welding shops often adjust their largest-capacity arc welders to about one-third of their capacity. For example, a 400-amp machine usually will be set at 100—125 amps.

Duty Cycle—More important than a welding machine's amperage is its *duty cycle*—the percentage of time the welder can be used at its rated output before it must "rest." This rest allows the machine to cool before resuming the weld. Exceed the duty cycle and the machine will gradually develop less than its rated output. A typical duty-cycle rating is 60% at 200 amps. This means the machine can be used for six out of 10 minutes while set at 200 amps. Duty cycle goes down when amperage is increased; it increases when amperage is reduced. A cost-efficient choice is a welder with a 90% or 100% duty cycle at 100—125 amps. But how do you determine

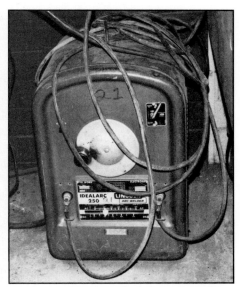

A-c/d-c welder is typical of many welding-school machines. It should perform for a long time without repairs.

Breven Finch inspects 220-volt a-c/d-c welder at Montgomery Ward store. Next to it is a 220-volt a-c welder. Both machines are ready to plug in to 220-volt service.

what the duty cycle of a welder is when it's adjusted to an amperage other than its rated setting? Here's how:

A welder's duty cycle at amperage settings other that its rated amperage is *inversely* proportional to the square of the new setting. In mathematical terms:

$$T_a = T_o(I_o/I_a)^2$$

Where:

I_o is rated amperage.
I_a is desired amperage.
T_o is rated duty cycle.
T_a is desired duty cycle.

In the above example, if the 200-amp machine is used at 125 amps, its new duty cycle is 60%(200 amps/125 amps)^2 = 154% duty cycle. The answer is 100% duty cycle because this figure cannot be exceeded. However, if amperage were increased to 180 amps, the new duty cycle would be 60%(200 amps/180 amps)^2 = 74% duty cycle, a substantial reduction.

When considering duty cycle, keep in mind that you won't be able to weld 100% of the time regardless of what the welder is capable of. You must stop to change rods, take a rest, change positions, reset the welder, or stop to weld another part or weld seam.

BUYING AN ARC WELDER

As was discussed in Chapter 3, never buy a welder without trying it out or having it demonstrated for you. This is particularly true with an arc welder. However, before heading out to "test drive" a welder, you should have a good idea of what type of machine will best fit your needs.

A-c or d-c?—If you plan to do occasional TIG welding later on, buy an a-c/d-c welder. But, if you just want to arc-weld with stick electrode, an a-c welder should be sufficient. Usually, the a-c-only welders have light-duty ratings and prices. Here's how to decide:

- For hobby welding a few hours a day, an a-c-only welder is fine.
- If you're a metal fabricator, and plan to weld a lot of mild-steel angle and plate, get a combina-

Typical a-c arc welder is usually run at 130-amps maximum, with up to 225 amps available. I've never welded *anything* at 225 amps. That doesn't mean you won't.

tion a-c/d-c machine.

- An auto bodyman fabricating many inches of weld a day needs a d-c wire-feed welder. See Chapter 11 for details on wire-feed welding.
- If you want to do high-quality welds for aircraft and race cars, get a TIG welder. See Chapter 9 for this information.
- If you plan to do *field welding*—no electric service nearby—consider a gasoline- or diesel-powered generator/alternator-type welder.

Portable d-c welder powered by gasoline engine can be used anywhere because it carries its own generator. Welder can be transported in light truck or trailer.

It's commonplace to find a portable welder hoisted high above a construction site. This advertises the welding contractor, discourages thieves and keeps vandals from tampering with the machine.

D-c welding with low-hydrogen rod such as E-7018 produces very smooth weld beads. Depending on the skill of the weldor, these welds can meet the strict X-ray inspection standards of the Atomic Energy Commission and oil-drilling industry. Low-hydrogen welding rod allows less hydrogen contamination in the weld bead, which helps avoid cracking.

D-c welding machines are usually more expensive than their a-c counterparts. They are used in schools and manufacturing plants requiring 100% duty cycle. They have to weld 10 minutes out of every 10 minutes.

D-c welding machines deliver a more *stable* arc because the *polarity* is not switching, resulting in a smoother weld bead and more uniform penetration. That's why d-c welders are used for more critical welds.

INSTALLING A 220-VOLT OR 440-VOLT WELDER

You don't have to pay much for a small a-c 225-amp *buzz-box* welder. However, if your workshop doesn't have an outlet to plug it into, it will take you a day or two and more than the cost of the welder to have an electrician wire it in. And, no, you cannot plug a 220-volt welder into an

A 220-volt welder plug won't plug into your electric stove or dryer outlet. But you can make an adapter. See page 22. Photo by Tom Monroe.

STOVE/CLOTHES DRYER OUTLET

ARC-WELDER 220V SOCKET

Note differences between receptacle for 220-volt welder and electric clothes dryer or stove. Check with electrician to make sure wires are of sufficient gage to operate arc welder before changing receptacles.

outlet meant for plugging in a clothes dryer or electric stove. The outlets are different, as shown above.

If you buy a 440-volt welder, it's possible that you'll have to rewire the building's main service box to obtain the power needed for your welder.

Now that you're aware of the possible electrical problems, don't be bashful about buying an arc welder. Once you overcome any

wiring problems, final setup is easy.

ARC-WELDING ROD

Refer to Chapter 12 for specifics on which fillers to practice with. Pick two or three kinds of welding rod. I suggest starting out with 5 lb of E-6011, E-6013 and E-7018.

Storing Arc-Welding Rod—All coated electrodes should be stored in a dry, warm atmosphere. I've seen it stored the following ways:

SUGGESTED ARC-WELDING ROD FOR PRACTICE	
ROD	**COMMENTS**
E-6011	Easy to use, but spatters a lot.
E-6013	Easy to use, has little spatter and produces average bead.
E-7018	Harder to use, but produces beautiful welds on clean metal.

Use one of these rods when making your first practice bead on steel.

Typical oilfield-pipe welding shop has overhead chain hoist for moving heavy pipe and positioning it for welding. Photo by Roger Finch.

Make welding as easy as possible. Here, my son, Rocky Finch, rolls pipe by hand on roller-equipped sawhorse as he welds with E-7018 rod.

Although tiring in this position, weldor holds rod straight out for accuracy. Pipe is rotated on rollers for better welding position. Arc glare is bright because heat is high. As additional passes make bead thicker, amperage is increased to increase penetration.

Coated electrodes must be kept dry. Large 480-volt three-phase oven will hold up to 400 lb of rod at 100–800F (38–427C). Small, 120 a-c/d-c or 240-volt a-c portable oven has 13-lb capacity. It can maintain the same temperatures. Photos courtesy Phoenix Products Company, Inc.

- In plastic pouches with the ends taped airtight and a couple of packages of *desiccant* crystals inside. The desiccant absorbs moisture. Although this method keeps the rod dry, the pouches are cumbersome to use when welding.
- In a metal tube with a tight-sealing lid and desiccant inside. This is the easiest-to-use method, but you'll have to make your own sealable metal can.
- In a 5-gal metal can, or *rod oven,* with a seal on the lid and a 60-watt light bulb inside. The light burns 24 hours a day. This method works OK, but keeping the 60-watt bulb lit continuously gets expensive.
- In a commercially built rod oven. Because of the expense, this method is practical only for welding shops using over 50 lb of rod a month.
- In an old refrigerator that does not cool anymore, with a light bulb burning inside continuously to keep the box warm and

moisture-free. This is for shops that keep several hundred pounds of welding rod on hand at all times.

When doing certified welding with E-7018 low-hydrogen welding rod, dry rod is very important. Weld quality is so important at nuclear power plants that if the

rod is exposed to the atmosphere more than eight hours, it must be thrown away. Although E-7018 rod can be reheated to drive out moisture absorbed from the air, it is never as easy to weld with as it was when fresh and dry.

Of course, all arc-welding rod should be kept dry. I've even seen E-6011 rod fail to maintain an arc because it was improperly stored for several weeks and, as a result, absorbed considerable moisture.

USEFUL EQUIPMENT

- Arc-welder cart. Every welding shop I've worked in had the arc welder on wheels—on a cart or even a trailer. The reason for this is basic. It isn't possible to bring all welding jobs to the machine. Instead, you'll have to take the machine to the job now and then.
- An arc welder should be supplemented with a gas-welding rig, including a cutting torch. Otherwise, you would have to do all your cutting with a hacksaw, metal shear or other metalworking tool. To give you an idea of how useful one can be, I use my gas-welding outfit 10 times more often than my arc welder.
- Hand grinder for dressing welds and beveling edges of metal plate prior to welding. A disc sander fitted with a cup-type stone or 9-in. grinder works well for this.
- Bench grinder. This type of grinder is handy for grinding small parts. Either mount the grinder on your workbench, fabricate a stand for your grinder, page 138, or buy one.
- Chipping hammer. This special hammer is necessary to remove slag from arc welds.
- Several wire brushes. Use a wire brush to clean off the slag after chipping.
- C-clamps. These are like having several extra sets of hands. C-clamps are a must for holding parts together or in position for welding.
- Marking Tools. Turn to Chapter 4 for information about these tools. Regardless of the type of welding or cutting you'll be doing, you'll need to accurately indicate where cuts and welds are to be made.
- Set of handtools for disassembly and assembly work.
- Spare weldor's helmet and lens for a helper to use or for a friend to watch with.
- First-aid kit with burn ointment. You'll need a first-aid kit eventually, even though you are super careful. It's almost impossible to work around hot metal without getting burned now and then.

Custom-fabricated adjustable jack stands support pipe assembly at a comfortable height for standing weldor. Photo by Roger Finch.

Weld bead at flange made with E-7018 rod has smooth, even ripples. This was a two-pass weld; *root pass* and *weld-out pass*. Photo by Roger Finch.

Arc welder can be a farmer's best friend when a tiller breaks at cultivating time. Photo courtesy Lincoln Electric Company.

PRACTICE

In this section, I describe how to arc weld using a-c welding machines. If you want to use a d-c machine to practice, read the section on polarity, page 83.

Scrap Metal—If you've been collecting pieces of scrap metal, you should have some pieces of 3/16- or 1/2-in. mild steel. Use a cutting torch to make 2 X 5-in. pieces on which to practice. This is a good size—large enough to work with, yet small enough to conserve your scrap pile.

Basic Practice Steps—As you did when learning to gas weld, learn the four basic types of welds before you begin a project. These are:

- Running a bead
- Butt weld
- T-weld
- Lap weld

For practice welding, you need a table with a 1/4- to 1/2-in.-thick, 2 X 3-ft steel top. In a pinch, you could simply lay a steel plate across two wooden stools or sawhorses. Later on, you could build an arc-welding and cutting table.

Miller Starfire a-c arc welder is portable, 100% duty-cycle machine available with gasoline or diesel power. Photo courtesy Miller Electric Mfg. Co.

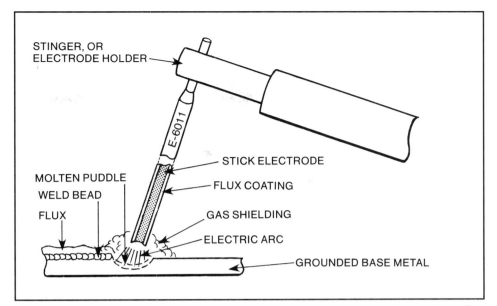

STINGER, OR ELECTRODE HOLDER

E-6011

MOLTEN PUDDLE
WELD BEAD
FLUX

STICK ELECTRODE
FLUX COATING
GAS SHIELDING
ELECTRIC ARC
GROUNDED BASE METAL

Arc welding generates high-temperature, gas-shielded *metal spray* to create molten puddle on base metal. Molten puddle solidifies as rod is moved along weld joint, leaving slag-coated weld bead.

Portable a-c arc welder being put to work mending steel fence.

Basic Arc-Welding Principles—

Before you strike the first arc, you should know what happens at the electrode tip. A 6000—10,000F (3320—5540C) temperature is generated by an electric arc between the electrode tip and the workpiece. The flux coating on the welding rod is heated to a gas and liquid. This shields the molten puddle from the atmosphere; thus the name *shielded metal-arc welding* (SMAW). The shield prevents the molten puddle from chemically reacting with and being contamina-

ted with atmospheric gases, causing hydrogen embrittlement, porosity and other bad effects.

As the weld puddle solidifies, the flux also solidifies, forming a coating on the weld bead and protecting it from the atmosphere as it cools. This resolidified flux—slag—which is glass-like, can then be chipped off to reveal the weld bead.

In the above drawing, you can see the arc-welding process with *stick* electrode. The arc-welding rod actually sprays molten metal into the molten puddle on the base metal.

Remember: Where you point an arc-welding rod is where the weld metal goes! The heat and sprayed metal come off the end of the rod like a spray gun! Point the rod where you want the weld bead!

Striking an Arc—I taught my wife how to strike an arc, run a bead, and actually weld something practical in less than two hours! A complete novice, it only took her five minutes to learn to strike and maintain an arc! Use the accompanying photos to follow each step and learn to arc weld quickly!

Ready the Welder—The first thing to do when getting ready to arc weld is to *ground the workpiece.*

Before you can do any arc welding, you must ground workpiece. Although you can lay workpiece on grounded table, direct ground such as this is more positive. *Make sure you are wearing a helmet and protective clothing. Warn any bystanders to "look away."*

You can't start an arc without a ground. Either connect the ground clamp directly to the work or to the metal table you'll be welding on. If you don't connect the ground clamp to a suitable ground, *you could become the ground!*

Once you've grounded the work, adjust the machine to 130 amps. Although this is a *hot* setting, you'll have an easier time learning how to strike and maintain an arc. Once you've learned to do these two things, you can readjust the machine to a lower

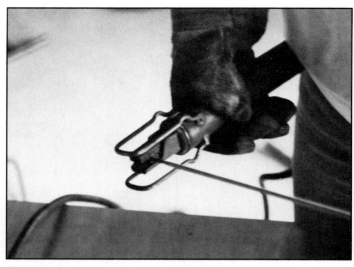

If you're just beginning, adjust machine to 130 amps. This "hot" setting makes it much easier to strike and maintain an arc. A more-advanced weldor can weld at lower amp settings.

Place base end of welding rod in holder like this.

amperage setting. Again, to make things easier, you'll need about five E-6011, 1/8-in. electrodes for practice.

The welding machine is now ready to be turned on. Make sure the *weldor* is. Regardless, don't do it while the working end of the welder—*electrode holder,* or *stinger*—is laying on the grounded table or workpiece. You may see some premature arcing.

Ready the Weldor—Don't turn on the welding machine until you've prepared yourself. You must be wearing the correct welding apparel: long-sleeve shirt, cuffless pants, high-top shoes, gloves and a welding hood. A leather apron is not necessary, but a good idea.

Turn On Machine—Turn the machine to ON. With the electrode in hand—your right hand if you're right-handed and vice versa if you're a lefty—squeeze it to open the jaws and insert the bare end of an electrode. Usually, there are grooves in an electrode-holder's jaws—for holding the rod at 90° to the holder, 45° forward, 45° backward and in-line with the holder. For now, position the rod so it's 90° to the holder as shown, above.

Strike an Arc—Weldors compare striking a welding arc to striking a match. However, a freshly lit match is immediately moved away

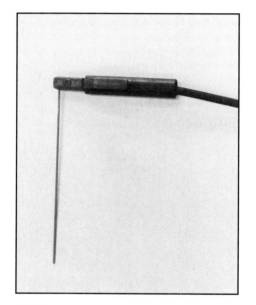

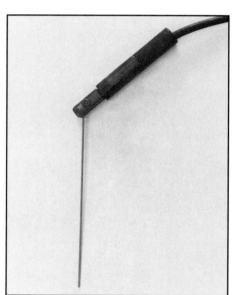

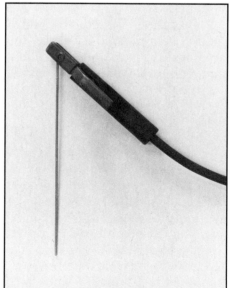

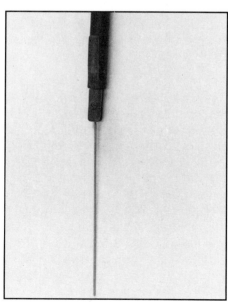

Rod can be postioned in rod holder differently for better access to weld seam. This is helpful when doing out-of-position welds. Photos by Tom Monroe.

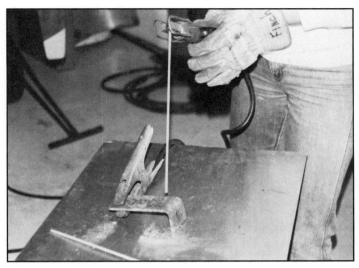

To start arc, hold rod holder with one hand and steady it with the other. Hold it close to workpiece and . . .

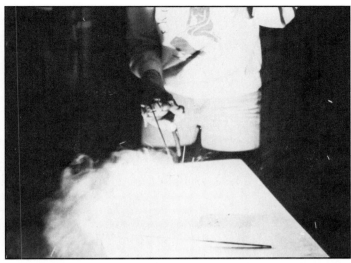

. . . nod your head so the helmet tilts down over your face. *Scratch* or *tickle* the metal with the rod. Once arc starts, keep rod close to work to maintain it. Don't stare at arc. Look at molten puddle. As rod melts, move rod holder closer to puddle to maintain arc.

from the striking surface. Not so with an arc welder. Once the arc is struck, you must keep the electrode tip near the work to maintain the arc. Instead of a match, the arc welder can now be compared to a sparkplug. If a sparkplug gap is excessive, it will not operate. The same holds true with the arc welder. Usually, you should maintain an electrode-tip-to-work gap of 1/8 — 1/4 in.

With these points in mind, let's get on with the business of arc-welding. With the stinger in both hands and welding hood or tinted shield flipped up—depending on the type of hood you have—hold the electrode tip about 1 in. from your workpiece. You are going to scratch or "tickle" the work with the rod to strike, or start, the arc.

With a mental picture of where the electrode tip is, nod your head so the helmet will fall down, covering your face and eyes. The next light—let's hope—will be the arc. Like writing with chalk on a blackboard, scratch the work with the electrode to start the arc. You can't just touch the electrode tip to the work to start the arc—the *tip must be moving.* Once started, keep the small gap just suggested to maintain the arc, and move along slowly. But, chances are you won't get this far on the first few tries.

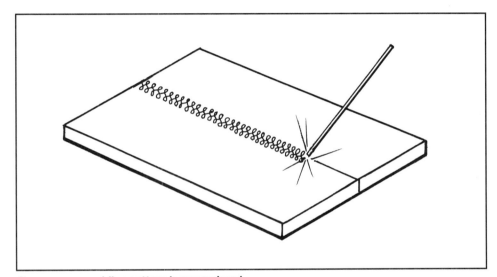

Common arc-welding pattern is weave bead.

If the rod sticks to the work—it happens to everyone—swing the stinger from side to side to break it loose. Do this quickly. Otherwise, the rod will get red hot and soft. If this happens and you can't break it loose, squeeze the holder to release the rod. You can then use pliers to work the rod loose. *Don't grab the hot electrode with your hands, even if you're wearing gloves!*

After learning to strike an arc, you'll have to maintain it. Do this by moving along while maintaining the correct gap. To do this, you'll have to move the holder closer to the workpiece as you move along.

The rod foreshortens as it melts to create the weld bead.

Once you've learned to strike an arc and maintain it, keep practicing while it's fresh in your mind. The object is to run a good weld bead.

Run the Bead—On a piece of scrap steel, practice running a bead until you're satisfied with its appearance. Use the pictures in this book as examples, or the welds on a car's trailer hitch or trailer. After welding a half-dozen or so acceptable beads on scrap steel, you should be able to try making a butt weld with two pieces of scrap metal.

Stand as steady as possible, feet spaced about 15-in. apart and one foot ahead of the other. Use two hands or lean against something to steady arc, if necessary.

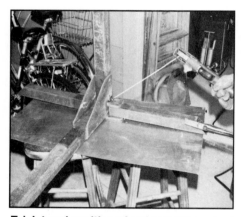

T-joint and position of rod and stinger just before striking arc. Remember to scratch metal with rod, then maintain 1/8-in. rod-to-work distance while welding. Midway through engine-stand project, page 139.

When butt-welding thick material, seam is beveled to obtain maximum penetration.

BEVEL

BUTT WELD

Items required for practicing a butt weld include two 2 X 5-in. steel plates, about 1/4-in. thick, and two E-6011 or E-6013 welding rods. The plates should be trimmed straight and even so they'll butt without any gaps. Place the two plates on your welding table. Butt them together, then tack-weld each end together.

As the first tack weld cools, the opposite end of the butt-weld seam will open up in a V-shape. Close the seam by supporting the back edge of one plate and tap on the edge of the other one with a small hammer. Once the gap is closed, tack-weld the other end of the seam. If the machine is still set at 130 amps, turn it down to 90 amps at this point in the practice. However, if the rod sticks in subsequent welding, turn up the machine to 130 amps again.

You're now ready to run your first butt-weld bead. Strike an arc at the right end of the seam, if right-handed, and slowly run a weld bead the length of the seam. The weld bead should be *centered* in the seam. After the weld is completed, let it cool for about five minutes in air. You can now pick up the metal with your pliers and stick it in water to complete the cooling.

Check for weld penetration. Look at the back side of the seam. If penetration is good, you should see signs of the weld puddle dropping out the bottom of the seam or extreme discoloration of the metal. If penetration appears to be insufficient, turn the heat up 15 amps and try again on two fresh strips.

Practice making butt welds until you think they look pretty good. To remove doubt as to the quality of your welds—a good looking weld isn't necessarily a quality weld—take your best samples to a technical-school welding shop. Ask for the instructor's opinion.

Test Weld—An easy way to test a butt weld is to clamp one side of the weld sample in a large vise, just below the weld seam. Hit the top of the side opposite the weld bead with a hammer. This will bend the metal toward the top side of the weld bead. If the weld is weak due to poor penetration, it will break through the back side of the weld seam.

Common mistakes for beginners are moving the rod too fast, resulting in poor penetration. Slow down! It's better to have too much weld bead than not enough.

For beginners, the best weld bead is obtained by moving the arc-welding rod at the same speed as the second hand on a wrist watch—about 3 in. per minute. As your skill improves, adjust travel speed to maintain a molten puddle.

T-WELD

Doing a T-weld with an arc welder requires that you *manipulate* the rod to avoid undercutting the vertical piece and to get good penetration in both pieces of metal. The tendency is to burn through the vertical piece and get insufficient penetration on the horizontal piece. This is the same problem encountered when doing

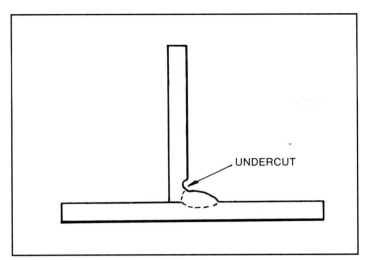

T-weld undercut is caused by vertical piece being overheated. Base metal then flows onto horizontal piece. Manipulate electrode so horizontal piece is heated most, but aim electrode at vertical piece.

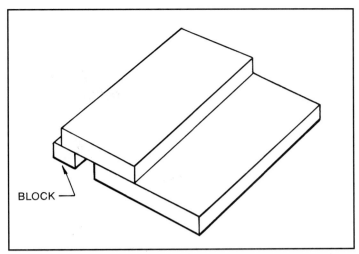

Block top workpiece level with bottom for doing lap weld. After welding lap joint, clamp one side in large vise and bend weld joint by hitting with hammer. Bending weld will show you how good or bad you're doing. If weld breaks, penetration is insufficient.

After two hours of arc welding, Breven folds engine stand to show how compact it can be made for storage. Stand is ready for sandblasting, painting, and set of casters.

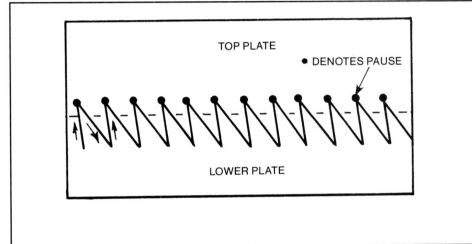

Weaving rod slightly makes better weld joint when lap-welding heavy steel plate.

a T-weld with oxyacetylene, page 52. The difference between the two types of welds is that you must now manipulate an electrode rather than a gas-welding tip.

To practice doing a T-weld, you need two 1/4-in.-thick steel plates that measure about 2 X 5 in. Place the two plates on your welding table so they form an upside-down T. Use a mechanical finger to hold the vertical section in place while you tack-weld each end. If the first tack weld causes the vertical plate to raise at the seam, tap it back in place with your hammer. Tack-weld the other end of the seam.

You'll need two sticks of E-6011 or E-6013 welding rod. If you're right-handed, strike an arc at the left side of the seam and start the bead. Apply about 70% of the heat to the flat part and 30% to the vertical part. This means you spend 70% of the time pointing the rod at the flat piece and 30% pointing the rod at the vertical piece. You'll have to swing the holder back and forth as you go while maintaining the arc gap and weld puddle. Have fun!

If you're left-handed, start at the right end of the seam and strike your arc on the *lower* piece. Starting on the upper piece would cause slag to be buried under the weld bead on the bottom piece. To time your stinger movements, count out loud, *one-two-three, one-two-three,* if it helps. You'll be constantly moving your rod-holding hand to accomplish this.

Again, when the weld is finished, let it cool. Chip off the slag and inspect the weld. If the vertical piece is undercut, suspect an excessive arc gap or rate of travel. Do it again on more scrap pieces.

Keep practicing until you're satisfied with the looks of your T-welds. Once you've mastered the problem of undercutting the vertical piece and can get sufficient penetration in the horizontal piece *with the same weld,* you can do a T-weld.

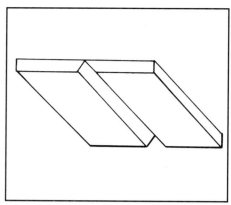

Workpiece positioned for doing overhead butt weld (left). Workpiece positioned for doing horizontal weld (right).

Horizontal weld is being made. Welding joint in any position except flat is *out-of-position welding.*

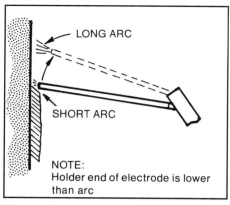

Keep electrode horizontal or pointed slightly upward when welding vertically up. As molten metal begins to deposit, move electrode tip 0.5—0.75-in. upward to allow puddle to solidify. Bring tip back to deposit metal. Continue this movement while you *only watch the molten puddle.*

LAP WELD

A lap weld is the most difficult type of weld to do with an a-c arc welder. You'll need two 3/8-in.-thick, 2 X 5-in. steel plates. Lay them on your welding table so one overlaps the other as shown. Tack-weld each end. E-6013 rod is best for lap welds, but E-6011 is OK in a pinch. As with a T-weld, the trick for doing successful lap welding is to prevent undercutting the top piece.

Again, strike the arc on the *lower* piece. Weave the rod in the motion shown on page 81. Be sure to pause slightly at the bottom each time you make a swing to prevent undercutting the top piece.

After you've completed running the lap-weld bead, cool the piece and chip off the slag. Inspect the weld for undercutting and penetration. Practice until you are making consistently good lap welds.

OUT-OF-POSITION ARC WELDING

Anything other than welding a seam that lays flat is *out-of-position* welding. This applies to all types of welds: butt, T or lap. And, out-of-position welding "will separate the men from the boys." Not only are they the most difficult to make, some are more difficult than others. Although I can give you tips on how to do each type of out-of-position weld, the only way to master each is with practice.

Regardless of your experience, always try to weld everything in the *flat position,* if possible. But many times it's impractical or impossible to reposition an item to make welding easier. So, you weld it where it is. Although there are "tricks" when doing each type of out-of-position weld, the number-one "trick" is to *point the rod where you want the puddle to go.* You're fighting gravity when welding out of position.

I'll never forget some out-of-position welding I once did while on the top rung of a 30-ft ladder. I was hanging upside down by my knees, welding the bottom side of a bracket on a light pole! While making that very uncomfortable weld, I said to myself, "Now, this is really out-of-position welding." That was several years ago, and I'm sure the bracket is still holding tight to that light pole.

If you have to arc-weld upside down, here's how to do it:

Overhead Welding—Use 1/8-in. E-6011 or E-6013 rod. Aim the rod so it points *almost* straight up, about 30° from vertical. Also, you should wear tight-fitting clothes. Sparks will fall on you.

After you start the arc, hold the electrode tip closer to the work than the 1/8-in. gap that's normal for 1/8-in. rod. I like to *push* the rod and puddle up against the base metal.

If you see a drip starting, push it back into the fresh weld bead, or if too big, fling it out of the puddle and start over. You may have to stop occasionally to let the weld bead cool. Usually, three or four seconds is enough cooling time. If you stop longer, chip out the slag to prevent slag inclusions in the weld.

Vertical & Horizontal Welding—For practicing vertical and horizontal welds, I recommend a 1/8-in. E-6013 rod. Other rods can be used. For instance, E-6011 is OK, but E-7018 is more difficult to use. Even though you may be using E-6013 rod, don't think vertical or horizontal welding is easy. This type of welding is the most difficult to perform and get a good weld.

For vertical welding up, point the rod about 15—25° upward from horizontal; 30—45° when welding down. If you weld at 90° to the work, the puddle will sag or fall out.

It is particularly important to watch the puddle, not the arc, when doing a vertical or horizontal weld. Otherwise, you won't be able to control the puddle. If it begins to sag, momentarily pull the rod back to lengthen the arc and stop depositing metal. This also cools the puddle slightly. Move back in to continue the weld. This in-and-out movement

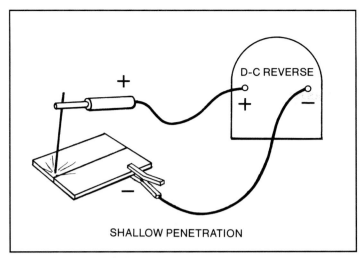

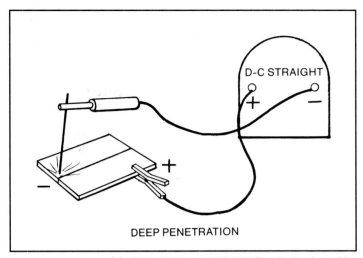

SHALLOW PENETRATION

DEEP PENETRATION

D-c welder set to (+), REVERSE, or POSITIVE polarity, has this circuitry. *Welding rod gets hotter* when machine is adjusted this way.

D-c welder set to (-), STRAIGHT, or NEGATIVE polarity, has this circuitry. *Workpiece gets hotter* when machine is adjusted this way.

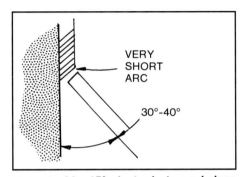

VERY SHORT ARC

30°-40°

Maintain 30—45° electrode-to-workpiece angle when welding vertically down. Move fast, otherwise slag will catch up with arc.

ARC-WELDING POLARITY

Metal	Polarity (a-c/d-c) in order of preference	Recommended Electrode
Stainless Steel	d-c Reverse (Positive) a-c	E-308-15, E-310-15 E-308-16, E-347-16
Bronze	d-c Reverse (Positive)	E-CuSn-C
Aluminum	d-c Reverse (Positive)	AL-43
Cast Iron	d-c Reverse (Positive)	ESt
High-Tensile Steel	d-c Reverse (Positive) a-c	E-7010-A1, E-8018-C3 E-7027-A1, E-8018-C1
Mild Steel	a-c d-c Reverse (Positive)	E-6011, E-7014, E-7018 E-6010, 5P, E-7018

is indicative of how to do such a weld, particularly *vertical up.*

Vertical-up welds—the puddle movement is up—are difficult to make because the puddle is below the arc. You can't use the rod to "hold" the puddle. One way to keep the puddle in position is to momentarily interrupt the arc, "freezing" the puddle so it stays in position. You can also momentarily *bury* the end of the rod in the puddle, also reducing heat.

Vertical-down welds—puddle movement is down—are easier to control because you can use the rod to push or hold the puddle in position as it tries to fall or drop out. However, penetration on a vertical-down weld is not as good as that for a vertical-up weld. For this reason, vertical-down welds

are not permitted when welding certain types of industrial pipe, such as high-pressure steam pipes in power plants.

Finally, don't forget to point the rod where you want the puddle to go. Although you've read it before, here it is again. All it takes is practice, practice, practice.

ARC WELDING WITH D-C CURRENT

Polarity—When welding with alternating current (a-c), you don't have to set *polarity*—direction of current flow. In the U.S., a-c constantly switches back and forth, 120 times a second, between positive and negative. A complete cycle occurs 60 times a second. In other countries, it may be 50 or 90 times per second. In direct-current

(d-c) welding, polarity makes a big difference. The above illustrations dramatize the effect of polarity in d-c welding.

Almost all d-c welding is done with *reverse polarity*—electrode is positive—because the welding rod gets hotter than the workpiece! Reverse polarity provides a steadier arc, and electrode-to-work metal transfer is smoother than with *straight polarity*—electrode is negative. It is easier to weld with a shorter arc and low amperage. Therefore, d-c is better for making out-of-position welds. Some electrodes can be used with reverse or straight polarity; these are called a-c/d-c electrodes. And, there are some that can be used only with straight polarity.

The above chart gives recom-

Old standby of welding industry is Lincoln Idealarc 250-amp a-c/d-c arc welder.

A-c arc welder and oxyacetylene torch were used for making mounts and brackets for Toronado-to-Corvair engine swap.

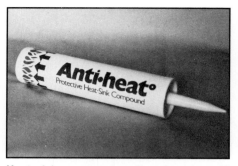

Heat-sink compound confines heat to weld area, minimizing discoloration or warpage of light-gage metal. Anti-heat is a product of Tempil Division, Big Three Industries, Inc. Photo by Tom Monroe.

mended polarity settings for various metals. Keep it near your welding machine so you can refer to it. You have the author's and publisher's permission to photocopy the chart for that purpose.

STOPPING & STARTING ARC WELDS

One thing common to all types of joints or arc-welding positions is the need to stop, then continue the bead. You may have "burned" the electrode to a stub and need a new one, need to change position, or rest yourself or the machine. Whatever the reason, you must get a smooth transition without voids at the end of one bead and the beginning of the next.

To get a good transition between two overlapping weld beads, do this: Stop and allow the weld bead to cool for about two minutes. With your chipping hammer, knock off the slag at the end of the bead. This is particularly important with deeply grooved weld joints. The slag will flow into the groove at the end of the weld, causing an *inclusion* or void in the weld if the weld bead is continued over it. Therefore, all slag must be removed completely.

Chip out the slag with the pointed end of a chipping hammer. Finish up by wire-brushing the end of the weld bead. When the seam is perfectly clean, continue welding. Overlap the welds by

striking the arc in the small crater at the end of the weld bead and immediately continue running the bead once a puddle forms. If done right, you should have difficulty seeing the transition between the two beads.

ARC-WELDING SHEET METAL

Welding sheet metal with an arc welder is difficult to learn because sheet metal is thin and easy to burn through. Here are some tricks to make it easier:

- Weld at low-amperage settings. Try 60—75 amps with 1/8-in. rod or 40—60 amps with 3/32-in. rod.
- Hold a very close arc. This keeps excess heat down long enough for the puddle to stick to the base metal.
- Stitch-weld or spot-weld, then tack and fill the gaps. This prevents local heat buildup and burning holes in the metal.
- Use lap welds, if possible. This thickens the metal, creating more *heat sink*—mass to absorb heat.
- If all else fails, use copper strips or heat-sink compound to back up the weld seam. The weld will not stick to copper; you can remove the strips after the weld has cooled.

ARC-WELDING 4130 STEEL

Often referred to as *chrome moly,* 4130 steel welds similar to

mild steel. However, the resulting weld is more prone to cracking after it cools because of 4130's "graininess." Here are some tips for a-c arc-welding 4130 steel:

- The larger the piece, the more important it is to preheat before welding. Always try to weld 4130 steel in 70F (21C) or higher temperatures, and preheat to 200—300F (93—149C) in the expected heat-affected zone. Use a temperature-indicating crayon or paint to monitor preheat temperature.
- Preheat with an oxyacetylene torch—rosebud tip if it's a large part—or heat the part in a kitchen oven if it fits. For huge parts such as nuclear power-plant reactors, an electric blanket is used for preheating.
- Always use E-7018 rod for welding 4130 steel.
- Make sure the base metal is clean and free of rust, paint and grease. Otherwise, you'll end up with a defective weld.
- Bevel the weld joints to get maximum weld penetration.
- Before taking a chance on ruining an expensive piece of 4130 tubing or whatever, practice on scrap.

Welding 4130 steel is discussed in more detail in the gas-welding chapter, page 54, because gas-welding is more suitable in most instances. For example, gas-welding is better for building a chrome-moly airplane fuselage or race-car-suspension parts. A gas torch is easier to manipulate around small parts and preheating is almost automatic.

ARC-WELDING ALUMINUM

Although uncommon, it is possible to arc-weld aluminum plate, aluminum castings and aluminum sheet with a d-c arc welder. The resulting weld bead will look rough compared to arc-welded steel. I've used it for welding large, thick aluminum plates, building up worn edges on aluminum pieces and welding 1/4-in. aluminum for toolboxes, barbeque grilles and shelf brackets.

As with steel, you should preheat 1/4-in.-and-thicker aluminum to 300—400F before arc welding. Expect a very bright arc, and a lot of noise and spatter when using aluminum arc-welding rod. The resulting arc-weld bead will be about 50% weaker than it would be if it was TIG welded.

ARC-WELDING STAINLESS STEEL

Although there are no special tricks to arc-welding stainless steel, don't expect beautiful welds. Stainless weld beads are not pretty unless they are *completely* protected from the atmosphere. The back side of the weld usually will appear black and rough.

Appearance of a stainless-steel weld can be improved by coating the back side of the seam with flux paste. This protects the seam from oxygen in the atmosphere, minimizing crystallization of the weld. Check the suppliers list on page 156 for stainless-steel back-shield paste.

The best welding processes for stainless steel are TIG and wire-feed (MIG). But if MIG or TIG are not available, you can do an acceptable job with an a-c arc welder. Select the correct stainless-steel rod from the chart, page 121. Again, there are no special methods necessary for arc-welding stainless steel. Preheating is not necessary.

Do the same as you would when welding any material for the first

When arc-welding near machined surfaces, area is coated with anti-spatter such as this spray-on compound from Weld-Aid Products. Photo by Tom Monroe.

time. Practice on scrap stainless before you try welding an actual part. If you don't have any stainless-steel scraps to practice on, you can use mild steel or 4130-steel scraps with a stainless-steel rod.

If you were wondering, yes, it is possible to weld mild steel, 4130 steel and stainless steel together in one assembly.

ARC-WELDING CAST IRON

You may encounter the need to arc-weld cast iron. Usually, this involves the repair of a cast-iron machine base, farm equipment, transmission case or engine part. The last time I welded cast iron was when a local high school asked me to build a chariot for a football-game halftime show. The students wanted it to look like Roman warriors coming into the arena. I was asked to weld late-model car axles to some cast-iron spoke wheels from a piece of old farm equipment.

I used *NIROD*—nickel-based welding rod—for doing this job. Or, I could've used Lincoln's *Ferroweld* or *Softweld* rod. The American Welding Society codes for these last two rods are ESt and ENI-CI.

It's hard to weld cast iron with-

out it cracking. The reason is its rigidity. When one small area is heated, causing it to expand, the unheated area resists. Unfortunately, the cooler area loses the battle because cast iron is much stronger in compression than in tension. Thus, the cooler area—in tension—cracks. This is why, it is extremely important to *thoroughly* preheat cast iron before welding it.

As you may suspect, welding cast iron requires a lot of patience. Start by heating the *entire* casting to 400—1200F (204—649C) before welding. Here's another application for temperature-indicating crayons or paint. And, only weld while the casting is hot. This is easy to do with small castings that fit into your kitchen oven, but large castings require a lot of heat. Because of this, it's standard practice for shops that weld cast iron to place a small natural-gas burner under the casting and heat it while it's being welded.

Another approach to welding cast iron—especially big castings that are impractical to preheat—is to arc-weld them at room temperature, but only 1/2-in.-long beads at a time, then stop. The 1/2-in. weld is chipped and allowed to cool for two or three minutes before another 1/2-in. weld is done. This allows the weld and heated area to "relax" as the heat is absorbed or dissipated into the casting. Some people recommend hammer-peening each short, fresh weld until it is cool. I prefer to not hammer-peen.

Brazing is Best—The best way to join cast iron is to *braze* it. And you can't do that with an arc welder. Get out your gas-welding torch. I've brazed a lot of small cast-iron pieces with success. The secret is to start by *V-notching* the crack or joint completely to the center from both sides of the part, or all the way through from one side. After you've done this, preheat the casting to about 350F (177C). The welding torch will help maintain heat as you braze. Read about brazing in Chapter 7.

TIG Welding

9

TIG welding is well suited for accurate welding jobs such as this prototype race-car front suspension. Photo by Tom Monroe.

Years ago, TIG welding was thought to be magic! It definitely had an aura about it. When I was asked if I would like to try my hand at TIG-welding aluminum, I responded with a big, "Yes!" The next day, I was welding patio screen-door frames for government housing. Nothing to it. It took less than one hour of help from a professional TIG weldor and about 15 hours of practice to get the hang of it. Afterward, I took the test for *certified aircraft weldor*—and passed!

Passing the certification test for TIG weldor was relatively easy. I was already an accomplished oxyacetylene weldor. The point is, you should become proficient at gas-welding before trying TIG. Gas-welding requires the same basic skills—controlling the puddle, moving the torch, dipping the rod, running the bead, etc. And, just like riding a bicycle, these are skills you won't forget as you apply them to TIG and other advanced forms of welding.

Name Calling—Many people will say, "Don't call it *TIG,* Call it *Heliarc."* Some even say, "call it *GTAW."*

Linde developed their trade name for TIG welding, *Heliarc,* from the words *helium arc welding.* Helium is an inert shielding-gas that envelops the weld puddle, keeping it free from atmospheric contamination. Today, helium has been replaced (largely) by argon, argon/hydrogen or argon/helium mixtures. However, the name *Heliarc* stuck in common usage, even when referring to TIG welders manufactured by other companies. Consequently, most weldors have to stop and think about what TIG means—*tungsten inert gas* or, more specifically, GTAW for *gas tungsten-arc weld.* It's important for you to understand that these three terms refer to the same welding process, not three different ones.

This section describes various TIG-welding setups, from the expensive ones to setups that'll get you by. I start with the very best. As with other welding processes,

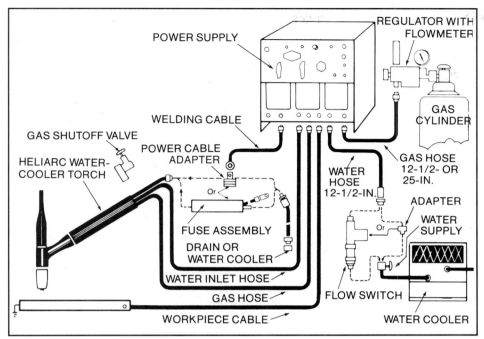

Basic water-cooled, a-c/d-c Heliarc setup is ideal for welding steel and aluminum. Note lack of high-frequency unit and amp-control foot pedal. Many race-car fabrication shops start with a setup such as this. Drawing courtesy Linde Welding Products.

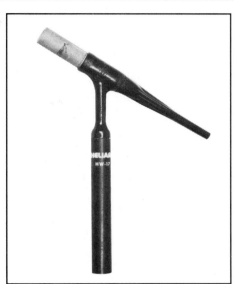

Linde Heliarc torch assembly. Photo courtesy Linde Welding Products.

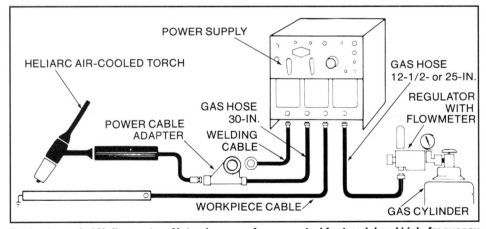

Basic air-cooled Heliarc setup. Note absence of amp-control foot pedal and high-frequency unit. Drawing courtesy Linde Welding Products.

you should know what you really need before you buy. Because a TIG outfit can be the most expensive of all welders, don't buy one until you have it explained and demonstrated to your satisfaction.

HOW TIG WORKS
TIG is the neatest, most precise and controllable of all hand-held welders. You could *almost* weld a razor blade to a boat anchor or shim stock to a crankshaft.

A small, pointed tungsten electrode—non-consumable—provides a concentrated high-temperature arc with pinpoint accuracy. You don't have to heat the whole area to start a puddle. Once the puddle starts, add filler just as you would with a gas welder.

Because of the TIG welder's high-heat concentration, but re-

duced heating of the workpiece, TIG is great for welding aluminum. Aluminum dissipates heat quickly, and the less heat absorbed by the aluminum, the better for the weld. If your workshop has a TIG machine, you could use it for most fusion-welding jobs except for rough ones such as building a race-car trailer. In fact, it can replace the gas-welding torch for all jobs except brazing or soldering. TIG welding has one major drawback—it's slow. So, for projects that don't require pretty welds but need to be done quickly, use an arc welder or MIG welder. MIG is covered in Chapter 11.

TIG COMPONENTS
Torch—Although more complex than a gas welder, the working end of the TIG welder is also called a torch. Instead of a flame, an electric arc is directed at the work to make and maintain the weld puddle. The arc occurs between the *tungsten*—a high melting-point, non-consumable

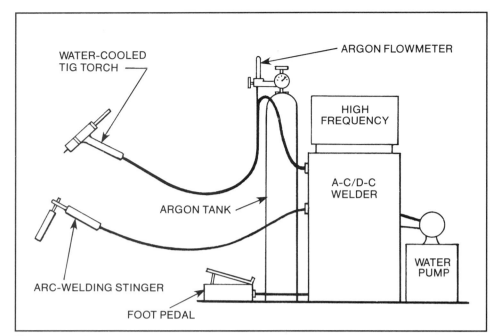

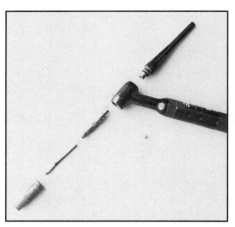

Disassembled water-cooled Linde TIG torch: From top, parts are long back cap, torch body, collet assembly, tungsten and nozzle. Photo by Tom Monroe.

First-rate TIG setup is self-contained with most equipment built into main power-supply cabinet.

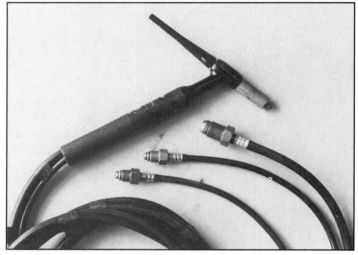

Water-cooled TIG cable is bulkier than air-cooled cable because it carries lines for water supply and return. Power cable is usually in water-return line. Note water and gas connections. Photo by Tom Monroe.

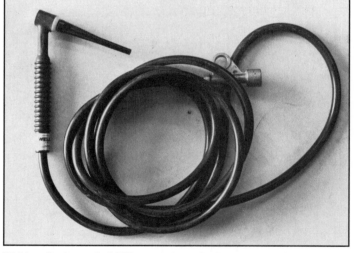

Weldcraft air-cooled TIG torch has single cable and connection. Photo by Tom Monroe.

electrode in the torch—and the workpiece. A *collet* in the torch clamps the tungsten so it can be adjusted in and out of the torch and retained in place.

Surrounding the tungsten is an open-ended *cup* that directs the shielding gas to the immediate area of the weld bead. Cups are ceramic because of the intense heat. Speaking of heat, TIG-welding torches must be cooled because of the close proximity to the weld puddle. It's not uncommon for a torch to get so hot that it's too uncomfortable to handle.

Torch *and* cable cooling are done with air or liquid—usually water. Water cooling is preferred by the serious user. However, an air-cooled torch is suitable for doing small jobs. Air cooling is not sufficient when welding for long periods or for welding thick material. How do you know when the torch gets too hot? Simple. It will burn your hand. The cable can get hot enough to melt the insulation!

Water, inert gas and electric power must be fed to a water-cooled torch. Consequently, the

torch has a cable for electric power and *three hoses*—one each for gas, water supply, and water return. Water is circulated through the torch and returned to the reservoir or dumped so the torch will receive a continuous supply of cool water.

Cups—The ceramic cup used on a TIG torch directs inert-gas flow over the weld puddle. A larger cup gives more gas coverage and improved weld quality. There are, however, times when a large cup such as a #10 will not fit into a corner or other tight area.

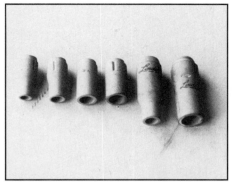

Assortment of ceramic cups, or *ceramics* or *nozzles,* as they are sometimes called. Photo by Tom Monroe.

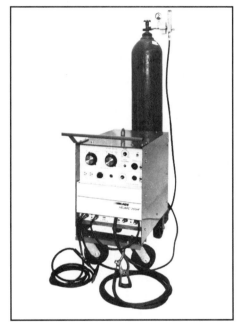

Complete TIG-welder setup by Linde. Photo courtesy Linde Welding Products.

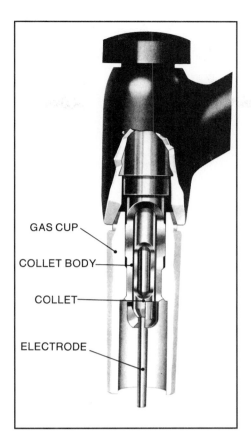

GAS CUP

COLLET BODY

COLLET

ELECTRODE

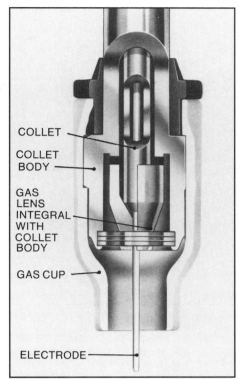

COLLET

COLLET BODY

GAS LENS INTEGRAL WITH COLLET BODY

GAS CUP

ELECTRODE

Standard TIG torch (left) and Linde torch equipped with gas lens. Note special cup used with gas lens. Otherwise, torches are the same. Photos courtesy Linde Welding Products.

CONVENTIONAL TORCH

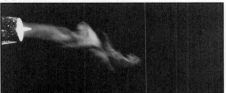

TORCH WITH GAS LENS

Gas lens smoothes gas flow, allowing tungsten to be extended for access to tight areas without loss of gas shielding. Note gas flow with conventional torch (left) and one with gas lens. Photos courtesy Linde Welding Products.

Consequently, a smaller cup must be used. Don't go smaller than a #4 cup. Gas coverage will be inadequate.

I use a #10 cup for flat seams, #8 cup for welding 1-in.-diameter tubing, engine mounts and race-car suspensions, and #6 or #4 for tight corners of aluminum air boxes and oil tanks.

Back Caps—These are used to clamp the collet and prevent the opposite end of the tungsten from arcing. Back caps are available in various lengths; short ones are used to weld in tight corners.

COMPLETE TIG SETUP

If money is no problem, several companies sell complete, first-class TIG-welding outfits. Such outfits will have built-in features to make welding easier. Here are some of these features:

Gas lens is a series of fine-mesh stainless-steel screens inside the torch-collet body that reduces shielding-gas turbulence. This focuses the shielding-gas stream, allowing the TIG weldor to extend the tungsten for working at greater nozzle-to-work distances such as needed in tight corners.

High Frequency is provided to start the arc by jumping a spark gap like a sparkplug. This is done by superimposing high voltage on the welding circuit. Otherwise, you would have to touch the tungsten to the base metal to start the arc. Touching the base metal is not desirable because the tungsten tip usually breaks off and ends up in the weld puddle. If it breaks off, the result is a contaminated weld bead. A broken tip also shortens the life of the tungsten. Touching the base metal may also contaminate the tungsten.

Popular Miller Gold Star 300 a-c/d-c power supply doubles as a TIG welder and arc welder. High-frequency unit is integral with power supply. Welder can be fitted with a unit that provides *slope* adjustment—gradual increase and decrease to and from weld current. *Downslope*—decrease in current—is used to eliminate cratering at end of weld bead. Photo courtesy Miller Electric Mfg. Co.

Foot-pedal amp control helps make better TIG welds, especially in aluminum. Only TIG machines with built-in remote amp control can use a foot-pedal amp control. Photo by Tom Monroe.

Although it is possible to weld steel without high frequency, it is required to weld aluminum or magnesium. Read on for more about this. Another point to consider: Welding with high frequency can ruin local TV or radio reception. Consequently, the use of high-frequency welders is regulated by the Federal Communications Commission (FCC) in the U.S. Follow the suggestions of the welding supplier or manu-

Foot pedal is positioned under table so weldor can operate it comfortably with right foot.

facturer when installing a high-frequency unit. Otherwise, you may become the most unpopular person in the neighborhood and on the FCC's most-wanted list.

Flowmeter allows you to monitor and regulate gas flow to the torch. Gas flow should be 10—25 cubic feet per hour (cfh). Any less would not *purge the weld*—displace all atmospheric contaminants around the puddle—allowing the weld to become contaminated. Much more than 20 cfh would be wasteful and could cool the weld too fast. Flowmeters have built-in pressure regulators set by the manufacturer. There are *flowgages* available that allow you to adjust both gas flow and pressure.

Cylinder is for storing a long-time supply of inert gas, usually argon.

Power Supply is similar to an arc-welder power supply. A TIG power supply may also include the following:

- **Foot-pedal amperage control** adjusts arc temperature and intensity. Rocking the pedal with your foot also starts and stops arc voltage. Pushing with your toe starts the arc and increases amperage for increased heat and weld penetration; pushing with your heel reduces amperage and stops the arc.

It can also be used for stick-electrode welding. When using a

Potential trouble: Welding booth is dark and cluttered with hoses and miscellaneous equipment. Take time to set up clean, neat weld area.

foot control, I set amperage about 20% higher than what I think is needed. This way, I get extra heat by pushing on the foot pedal. I don't have to stop to readjust the power-supply amperage control. Note: Stick welding with a foot pedal enables you to "back off" the amps at thin spots. Likewise, you can add amps where extra heat is needed. And stopping the weld with the foot pedal allows you to reduce heat at the edge of the workpiece and avoid a flat, thin bead common with conventional stick welding.

- **Solenoids** built into welder cabinet start and stop gas flow and cooling water. These solenoids self-time water flow to cool the torch and give timed *post-purge* of gas so the hot weld bead is not contaminated while cooling. Consequently, water and gas flow continue for a short time, even though the arc has been stopped. To take advantage of the post-purge gas, the torch must be held for a moment over the end of the weld bead after the arc has been stopped.

- **High-frequency power supply** has a cleaning feature required for TIG-welding aluminum and magnesium. Its continuous spark actually clears the puddle area by providing an electric-field shield over the weld bead. This is in addition to the feature

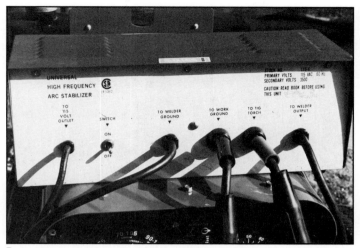

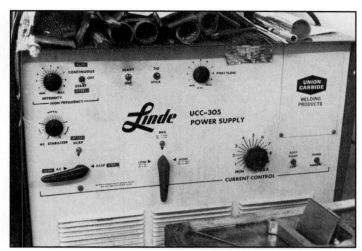

Box sitting atop portable Miller welding machine is separate high-frequency arc stabilizer. High frequency is necessary for TIG-welding aluminum or magnesium, and makes welding stainless steel easier. Miller power-source wire and switch connections from left to right: voltage control, on-off switch, welder ground, work ground, TIG torch and welder output. High-frequency arc stabilizer can be adapted to any 220-volt a-c or d-c arc welder. Also required is argon cylinder with flowmeter, TIG torch and cable. Photo courtesy American Steel and Welding Service.

Linde control panel looks different than Miller, page 90, however basics are the same. Once you get accustomed to your machine, others seem foreign. Photo by Tom Monroe.

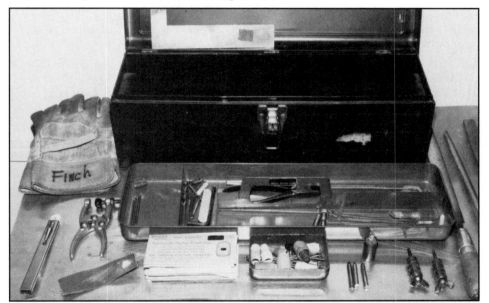

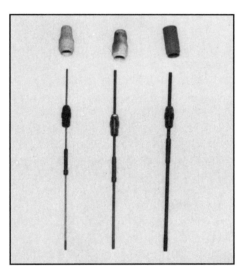

You don't need much equipment in your TIG toolbox, but gloves, extra tungstens and torch parts are necessary. They should be on hand if you do much TIG welding. Kit also includes files, marker and clamp.

Three tungstens, collets, chucks and cups most often used are, from left to right: 1/16 in. with #4 cup; 3/32 in. with #6 cup; and 1/8 in. with #8 cup. I usually grind tungsten in half to fit in a short torch.

that allows the arc to be started without the need to touch the tungsten electrode to the base metal. High frequency is shut off once the arc is established when d-c current is used; it continues with a-c current. A separate switch provides for start high-frequency or continuous high-frequency.

• **Range and polarity switches** select a-c, d-c straight-on or d-c reverse polarity for any kind of welding.

• **115-volt a-c receptacle** accepts work light, extension cord or other accessories.

• **Portable base or cart** contains necessary items such as gas bottle, water reservoir and foot-control pedal.

• **Water reservoir** is a nice extra because it allows continuous or high-amperage welding without danger of overheating the torch. A good water reservoir holds about 5 gal and includes a motor-drive pump. The water

should be mixed with 50% ethylene-glycol antifreeze if below-freezing temperatures are expected. If you use antifreeze, change it periodically. Antifreeze degrades and gives off a noxious odor when stale.

Necessary Extras—To supplement the deluxe TIG setup, you'll need a good arc-welding helmet, gloves and a small toolbox with trays for collets, chucks, cups, extra helmet lens and tungsten electrodes. I have two small trays

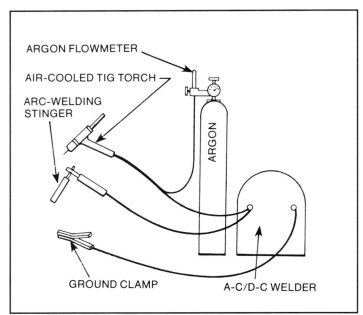

For TIG-welding steel and stainless steel, this bare-bones setup will suffice. You can even arc-weld with it until you can afford the argon and torch pieces.

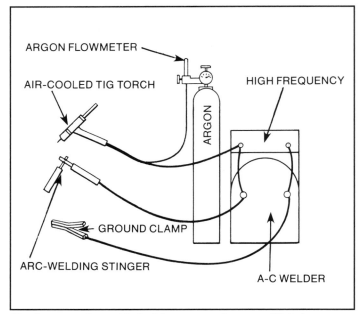

Aluminum can be welded with an a-c-only setup. High-frequency power supply with foot-pedal amp control should be used to do certified quality welding.

for tungsten—one for dull electrodes and one for sharp electrodes. When the tray of sharp tungstens is depleted, I sharpen the dull ones and transfer them to the *sharp* tray. This saves time when I need to change electrodes.

In the toolbox, keep several pins, clamps and setup fixtures handy. An important tool is a 6 X 12-in. framing square. If you recall from the gas-welding chapter, the framing square is useful for setting up parts and checking for possible warpage. Most things we weld are square to something, so a square comes in handy.

- **Files** of various sizes and descriptions. As discussed in the gas-welding chapter, you should have a coarse round file, a coarse half-round file and a flat mill file.
- **Bench grinder** with fine- and medium-abrasive wheels. Use these for sharpening tungsten and fitting small parts.
- **Belt sander and disc sander** for fine-dressing tungsten and fitting parts to be welded.
- Other niceties include a comfortable chair, clean welding table and good lighting. Mount a fluorescent light over your welding table. Two or three portable flood lights for welding

such things as engine mounts, race-car frames, trailers or airplane fuselages may come in handy. A clean, well-lit welding area is safer and will result in better welds.

MINIMUM TIG SETUPS

D-c Conversion—If you already have a d-c arc welder, you can convert it to a TIG welder by adding an air-cooled torch, and an argon flowmeter/flowgage and tank. This will restrict your TIG-welding to steel and stainless steel because of the lack of a high-frequency unit. Because of this, you'll also have to start an arc on the base metal or a copper *strike plate*. A strike plate makes arc-starting easier and reduces tungsten contamination from the base metal. Also, copper won't break the tungsten as easily as steel or aluminum, but it's not as good as high-frequency arc-starting. Without a foot-operated amperage control, you must stop welding to make amperage adjustments at the machine.

A-c Conversion—If you have an a-c arc welder, you can fit a high-frequency unit to it—for about the cost of your a-c machine—a torch, hoses and regulator, and argon gas cylinder. This will allow you to

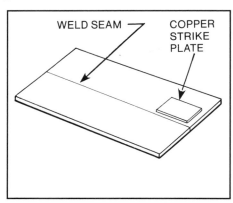

For TIG welding without high-frequency starting, use small copper strike plate for starting arc. Unlike steel, copper will not cause tungsten to break off when starting.

TIG-weld *aluminum*. Again, you won't have a foot-operated amp control. You'll also have to touch the tungsten to the base metal or strike plate to start the arc.

Whichever TIG setup you choose, you'll be able to weld things you never thought possible, such as welding thin parts to thick parts or complicated assemblies without burning them up. But mainly, TIG is so "squeaky" clean that you can weld in your Sunday clothes and not get dirty. Although I don't recommend this, I've actually welded airplane parts in a suit, white shirt and tie!

Although slow compared to arc and MIG welding, TIG welding is considered the most precise method of shop welding. Consequently, it's used to repair bad welds where accuracy counts.

RECOMMENDED TIG SPARE PARTS

To avoid making too many trips to the welding supply store, keep a supply of TIG-welder parts in your toolbox. At right is a shopping list.

USING A TIG WELDER

To refresh your memory, an electric arc is generated between a *non-consumable* tungsten electrode and the base metal in TIG welding. *Non-consumable* means that the electrode is not intentionally melted into the weld puddle as in conventional arc welding. The tungsten will, however, erode and become contaminated in use. Consequently, it will also be ground away as you dress it to the desired point again and again.

The type of metal being welded determines the tungsten-tip shape required. For instance, the sharp point needed to weld steel confines the arc to a smaller area, resulting in more concentrated heat at the weld seam. But, a crayon-shaped point is used to weld aluminum because that metal dissipates heat more quickly and needs more area heated at the weld seam. This is done with the resulting broad arc. More on dressing tungsten tips later.

As the arc heats a molten puddle on the base metal, dip the filler rod into it as you would if gas welding. The inert gas from the torch shields the puddle from atmospheric contamination.

To ensure that shielding gas covers the weld while solidifying, keep the torch over the bead after completion until the purge gas stops. The timer on the TIG machine usually provides 5—6 seconds of gas flow after the torch is off. This keeps 4130 steel from cracking and helps prevent crystallization in stainless steel. Also, titanium demands post-purge gas to prevent atmospheric contami-

SPARE PARTS LIST	
Number	**Description**
3	#10 ceramic cups
3	#8 ceramic cups
6	#6 ceramic cups
3	#4 ceramic cups
4	1/8-in. tungsten, 2% *thoriated (for steel)
4	1/8-in. tungsten, pure (for aluminum)
4	3/32-in. tungsten, 2% *thoriated (for steel)
4	3/32-in. tungsten, pure (for aluminum)
4	1/16-in. tungsten, 2% *thoriated (for steel)
4	1/16-in. tungsten, pure (for aluminum)
2	1/8-in. collets
2	1/8-in. chucks
2	3/32-in. collets
2	3/32-in. chucks
4	1/16-in. collets
4	1/16-in. chucks
1	stubby back cap
1	2-in. back cap
2	small stainless-steel wire brushes.

*Thoriated means tungsten includes thorium alloy for making it easier to start an arc. Unfortunately, thorium also can contaminate aluminum and magnesium, so pure tungsten should be used for these applications.

Reduce trips to the welding-supply store by keeping a supply of TIG parts in your toolbox.

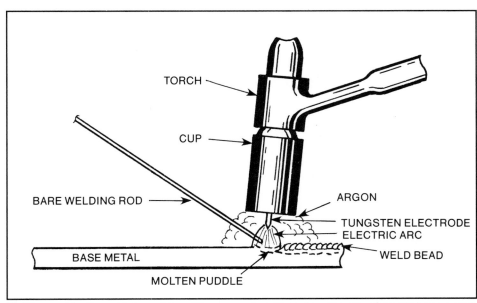

Welding rod is dipped into weld puddle as needed. Tungsten should not touch base metal, weld puddle or filler rod.

nation of the cooling bead.
Polarity Settings—Most d-c TIG welding is done on straight polarity—electrode is negative.

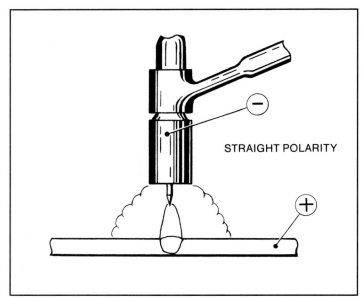

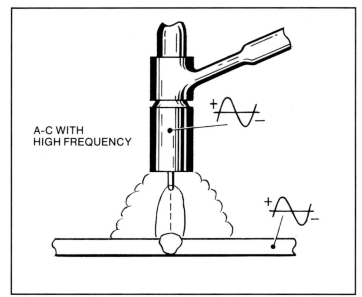

Set TIG welder to d-c current, straight polarity for welding mild steel, 4130 steel, stainless steel and titanium.

Set TIG welder to a-c current for welding aluminum and magnesium.

There is no polarity with a-c current. A-c current constantly changes direction. There are 60 complete cycles a second in the U.S. A-c current frequency is 50 cycles per second in Europe and 90 cycles per second in other parts of the world.

When welding most mild steel, stainless steel, 4130 steel and titanium, set your machine for *d-c current, straight polarity*. This concentrates most of the heat at the work—about 70%. However, when welding aluminum or magnesium, use a-c current. The reason for this is the oxide-cleaning feature of reversing current—very important when welding these non-ferrous metals.

As for TIG-welding with *d-c current, reverse polarity*—electrode is positive—you'll rarely have the occasion. The tungsten may melt before the base metal because about 70% of the heat is at the electrode, not at the work. Consequently, penetration of the base metal is poor. However, d-c current, positive polarity does have a useful application. It is best used for welding very thin sheet metal—not aluminum or magnesium. Shallow penetration then becomes an advantage.

The drawing on page 95 shows a typical fully equipped TIG-

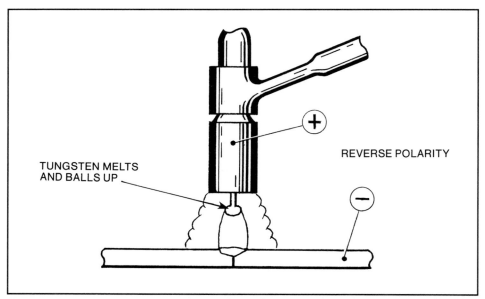

D-c current, reverse polarity is a seldom-used TIG setting because tungsten melts before base metal. Higher heat is at electrode, not work. Therefore, its only application is for welding thin-gage metal.

machine control panel. The settings on your welder might look different, but d-c straight polarity or a-c polarity should be as shown.

Tungsten Sharpening—I coarse-sharpen tungsten electrodes on a small bench grinder, then finish dressing them on a power sander with 120-grit paper. The grinder removes metal fast and the sander does a good dressing job. Remember to grind tungsten slightly blunt for welding aluminum and sharp for welding steel, stainless steel

and titanium.

A sharp tungsten tip is best for welding steel because it provides for better control and concentrated heat. But, because the heat for welding aluminum needs to be more-evenly distributed and a-c current melts the sharp electrode point—tungsten/base-metal heat distribution is about 50/50 on a-c—a sharp tip would contaminate the weld. The blunt tip cures both of these problems. It *scatters* the arc, distributing the heat, and

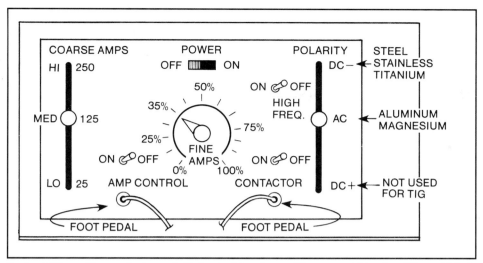

Actual Miller TIG-welder panel shown at left. Photo by Tom Monroe.

Typical TIG-welder control panel: Refer to this when setting up machine for different kinds of metal and welding methods.

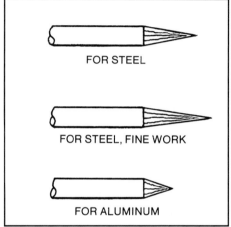

FOR STEEL

FOR STEEL, FINE WORK

FOR ALUMINUM

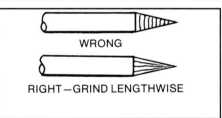

WRONG

RIGHT—GRIND LENGTHWISE

Shape of tungsten electrode is important. Sharpen tungsten to look like a pencil for welding steel. Sharpen to needle point for fine work. Sharpen tungsten to crayon-like point for welding aluminum. Always grind tungsten *lengthwise.*

it doesn't melt.

Don't go to the trouble of *balling,* or rounding the tip of, the tungsten for aluminum welding as some weldors do. It will ball itself in about two seconds of welding.

Sharpen tungsten lengthwise using a series of straight cuts toward the tip. *Never sharpen tungsten by rotating it against a grinder.* Sharpening a tungsten in this manner will result in a poorly controlled arc pattern. See drawings at right.

Install & Adjust Tungsten— Loosen the collet on the back of the torch to free the tungsten. This will allow it to slide in and out. Adjust the tungsten tip so it projects about 1/8—1/4-in. past the tip of the cup.

TIG-WELDING TIPS

To make things easier, read the following before you start TIG welding:

- I know you've heard this one before, but here it is again: Master gas welding before you attempt TIG welding.
- Clean the base metal as if you were going to eat off it. Seriously, you can't get the metal too clean for TIG welding. There is no flux to float off impurities.
- Cut the welding rod into 18-in.-long pieces. This usually

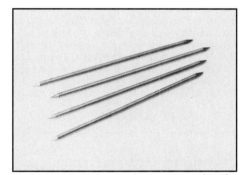

Keep extra sharpened tungstens handy. Sharpened at each end, tungsten can be turned around in torch so welding can continue. This eliminates need to stop to sharpen tungsten or replace it. Tungstens, cups and brushes for welding steel, aluminum and titanium must be kept separate to avoid contamination. Photo by Tom Monroe.

means cutting the 36-in.-long rods in half. Shorter pieces are easier to use.

- Tungsten diameter should be about half base-metal thickness. For example, use 1/16-in.-diameter tungsten to weld 1/8-in.-thick metal.
- Cup size should be as large as possible without restricting access to the weld. For instance, you'll have to use a smaller cup to weld in tight corners. Use a #8, #10 or #12 cup for flat steel and where access is good; #4 or #6 cup for corners and where access is poor.
- Clean the welding rod before you start welding. Use MEK or alcohol on a clean, white cloth.

Even dust contaminates the weld.

- Make sure the lighting is good because the light from the arc is less intense than in other welding. Use a clean, #10 helmet lens for most TIG welding.
- Do not allow even the slightest breeze or draft in the weld area. Cool air will crack a TIG weld because the heat-affected area is smaller and more sensitive to rapid cooling. A breeze can also blow away the shielding gas.

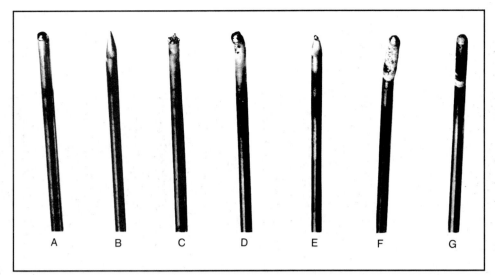

Tungsten looks differently when used on different materials under varying conditions: A, pure tungsten developed ball end when used with a-c current on aluminum. B is 2% thoriated tungsten ground to taper used with d-c current. C is 2% thoriated tungsten used with a-c current on aluminum. Tip developed several ball-shaped projections rather than ball end, A. D, pure tungsten used with a-c current on aluminum, was used with current above rated capacity. Overheating caused ball to droop to one side. Continued use would have caused end to drop into molten puddle. E, pure tungsten was tapered to point and used on d-c, straight polarity. Pure tungsten should not be ground to a sharp point because it will melt when arc is established. F was contaminated when filler rod touched tungsten. To correct, contaminated area must be broken off and tungsten reshaped. G did not have sufficient gas *post flow.* Oxidation of tungsten occured because electrode wasn't cooled sufficiently by shielding gas before coming into contact with atmosphere. Increased post-flow time should correct problem. Photo courtesy Miller Electric Mfg. Co.

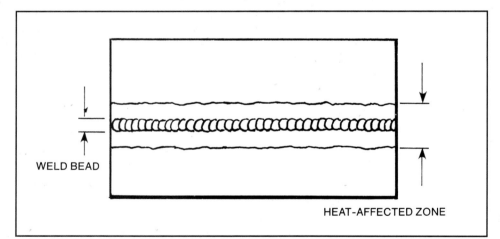

WELD BEAD

HEAT-AFFECTED ZONE

Heat-affected area of TIG-weld seam is less than that for most other types of welding. Because of high temperature differential and small heat-affected area, avoid TIG welding in a draft.

- Always shield the TIG-weld light so you won't burn someone's eyes. TIG appears to give off less light, but its ultraviolet radiation is just as dangerous.
- Sit down while welding, if possible. Be comfortable. Take advantage of the fact that TIG welding generates no sparks to fly into your lap.
- Always tack-weld parts before running your final bead.
- Before making a critical TIG weld, try your procedure on a test specimen first.
- As in oxyacetylene welding, make a molten puddle, then dip filler rod into it. Don't try to melt the rod with the arc. A *cold* weld, with poor penetration, results if you try to drop molten rod into the seam.

- Never touch the hot tungsten to the puddle or filler rod. Capillary action causes molten metal to *wick up*—flow up—the tungsten, contaminating it. The weld is also contaminated as the wicked metal oxidizes, and boils off the tungsten, blowing oxidized metal into the puddle. If this happens, stop welding immediately and grind the wicked metal off the tungsten. This is why you'll use more tungsten now than after you have more experience.
- When—not if—you burn a hole through the base metal, completely let up on the pedal or break the arc. Let the puddle cool before continuing.

Following are specifics on welding popular metals. Suggestions are given on how to set up the machine and provide for additional shielding, if needed. Except for magnesium, refer to Chapter 12 for information on specific alloys and the respective filler rods to use. Magnesium rods are covered in this chapter, page 103.

TIG-WELDING STEEL

Because mild steel and chrome moly are TIG-welded with the same process—machine and filler material—you can weld mild steel to 4130 or vice versa. For welding either, set up the machine by using the following procedure:

- Use steel welding rod #7052 or equivalent. See chart, page 122, for details.
- Use d-c straight polarity—negative electrode.
- Set high-frequency switch to START—if so equipped.
- Adjust gas flow to 20 cfh.
- Set water-cooling switch to ON—if so equipped.
- Set coarse- and fine-amp controls—coarse to 75 amps, fine to 80% for 0.063-in. steel. This gives 75 X 0.80 = 60 amps with the pedal all the way down. For thick metal, try these settings: coarse to 125 amps and fine to 35%, or 44 amps with full pedal. If this doesn't provide

enough heat, increase the fine-amp setting. Go 50%, 80% or whatever *heat* you need. The reason for the lower fine-amp setting for thicker material is to increase duty cycle. Regardless of the fine-amp setting, final amperage at the electrode is controlled by you at the pedal.

• Turn on contactor and amp-control switches to remote foot pedal—if so equipped. Contactor switch turns "on" the foot pedal.

• Turn on machine.

Ready to Weld—The torch should be in your right hand and rod in your left hand if you're right-handed. Switch hands if you're left-handed. To begin, rest the corner of the cup on the work at about 45°. Don't allow the tungsten to touch the work. Later on, you should be able to judge a 1/4- to 1/8-in. arc gap without touching the cup to the work.

Get comfortable. Rest your arms—covered with long shirt sleeves—on the welding table or workpiece. Assume the most comfortable position that allows you to operate the foot pedal. Never weld with your arms unsupported. If possible, lean your shoulder against something steady.

Keep the welding rod away from the arc until you have a molten puddle. *After* it has started, dip the rod in and out of the puddle as you move along the weld seam. *Do not* try to melt the rod with the arc. Let the molten puddle do the melting. Dipping the rod into the puddle cools the puddle slightly. So, the rhythmic in and out motion of the rod maintains a constant puddle temperature. If dipping cools the puddle too much, compensate by increasing heat slightly.

Tip: This one is worth repeating, particularly for doing TIG welding. Drill a #40 hole in tubing when welding it closed. This will keep the puddle from blowing out onto the tungsten as you close the seam. Hot air expanding inside the tubing causes this.

Beautiful workmanship: Peter Hoey's Christen Eagle biplane fuselage is TIG welded. To do such a project, you must observe correct welding practices. Tips from the experts are also helpful.

Start TIG-welding by resting corner of ceramic torch cup on metal. This keeps tungsten from touching workpiece. With experience, you'll be able to judge correct arc-to-work distance. Steady yourself by resting both arms on welding table or workpiece. If left-handed, hold torch in your left hand; or vice versa. And, wear long-sleeve shirt.

Perfect TIG weld in 4130 steel tubing should look like closely spaced fish scales.

TIG-welded aircraft engine mount looks better than gas-welded one because heat-affected area is smaller.

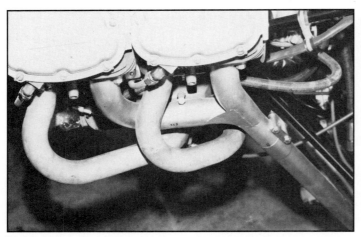

Stainless-steel exhaust headers should be *back-gas* purged before and during welding to prevent atmospheric contamination.

TIG-welded titanium exhaust headers on stock-block Indy Car: Titanium is highly susceptible to atmospheric contamination, thus back-gas purging is a must. This is especially critical with structural components. Photo by Tom Monroe.

Drill small hole in non-stressed area of tube that will be closed with end cap. This prevents puddle from blowing out as weld is completed. Photo by Ron Fournier.

TIG-WELDING STAINLESS STEEL

TIG-welding stainless steel is similar to welding mild steel and 4130 steel. However, you'll need some extra items: an extra bottle of inert gas, extra flowmeter and about 20 ft of argon hose. This assembly of items will be used to *purge*, or shield, the *back side* of the weld.

Back-Gas Purging—As mentioned earlier, purging is the process of displacing all atmospheric gases and replacing them with an inert shielding gas such as argon. I usually set purge-gas flow the same as or about 25% more than torch flow.

The reason for using *back-gas* purge is that molten stainless crystallizes if it's exposed to air. *Sugar,* or crystallization, on the back side of the weld would weaken the weld and base metal considerably.

It's easy to back-gas stainless tubing such as an engine exhaust pipe. You simply fill the tube with gas. Cap both ends of the tube with masking tape, punch holes in the tape, stick the *purge hose* in one end and a short piece of open tubing in the other end to exhaust the argon. At 15—20 cfh gas flow, the average exhaust pipe can be purged of air in 4—5 min. Larger pieces, such as long, 12-in.-ID tubing, take more time to purge.

Flat stainless plate is harder to purge. But, you can build a cardboard-and-masking-tape cover over most seams to act as a purge shield. Simply enclose the back side of the weld seam with craft-type cardboard and masking tape. Shape the cardboard so it straddles, but doesn't touch, the weld seam. Burn a large hole in the cardboard and the purge gas will be lost. Cap the ends with masking tape and insert the purge and exhaust hoses. The accompanying drawing illustrates how to make a back-gas shield for flat plate.

TIG-WELDING TITANIUM

Although most alloys of this expensive, lightweight metal can be TIG-welded with the same basic machine setup as mild steel and 4130 steel, it requires a more elaborate shielding-gas apparatus than stainless steel. As a result, some titanium alloys are not weldable at all. This is because of titanium's extreme sensitivity to contamination. Hot titanium reacts with the atmosphere and dissimilar metals, causing weld embrittlement. This contamination is serious if carbon, oxygen or nitrogen is present in sufficient quantities. In the solid state, as in a weld heated above 1200F (649C), titanium absorbs oxygen and nitrogen from the air.

An argon-gas shield must cover both sides of a weld seam while titanium is at the 3263F (1795C) molten stage and all the way *down to* 800F (427C). Otherwise, embrittlement from contamination and resulting cracking will occur.

Make a device to provide a *trailing shield* to cover the weld until it cools to below 800F (427C). Usually, after the trailing shield passes over the completed weld, the titanium has cooled sufficiently. You can also use temperature-indicating crayon or paint to be sure. But, if you do, be careful not to contaminate the weld with the crayon or paint. See

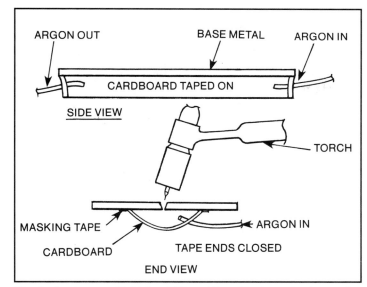

Simple back-gas shield for TIG-welding titanium can be made from cardboard. Masking tape is used to hold it to workpiece.

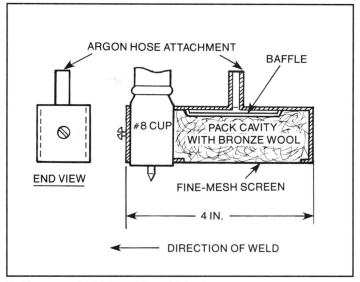

Trailing-gas shield for TIG-welding titanium protects completed weld as it cools. Shield must fit tightly to cup. See following drawing for optional design.

the drawings of trailing-shield devices on this page.

Gas Chamber—Because it is extremely reactive with nitrogen, oxygen and hydrogen, the best place to weld titanium is in a total inert atmosphere such as that in outer space. So, if you can get a ride on the space shuttle to do your welding, great. Otherwise, the next best thing is a *gas chamber*. Such a chamber looks similar to an incubator for new-

born babies or a bead-blasting cabinet. Put the part to be welded inside the chamber, close it, then put your hands into built-in gloves to do the welding. The chamber is completely purged with argon so no air exists. Much purer titanium welds will result.

TIG-WELDING ALUMINUM

TIG-welding aluminum is slightly different than welding steel. Machine settings are different. And,

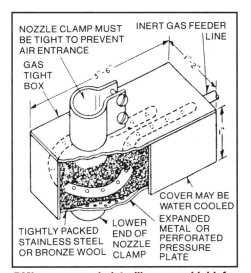

RMI recommended trailing-gas shield for TIG-welding titanium. Drawing courtesy RMI Company.

as you may remember from the gas-welding chapter, aluminum doesn't change color as it forms a puddle—it gets shiny instead. However, unlike gas welding, flux isn't needed to TIG-weld aluminum. In fact, flux would really mess up the weld *and* your TIG welder. The following tips will help you TIG-weld aluminum:

● The weld-seam area should be as clean as possible, *and* it should be free of aluminum oxide. Remove the oxide *immediately* before you weld—not a week or two before—by mechanical or chemical cleaning.

GAS SHIELDING FOR TITANIUM WELDING

Material Thickness (in.)	Torch Argon Flow (cfh)	Trailing Shield Argon Flow (cfh)	Back Gas Helium or Argon Flow (cfh)
0.030	15	15	3
0.060	15	20	4
0.090	20	20	4
0.125	20	30	5

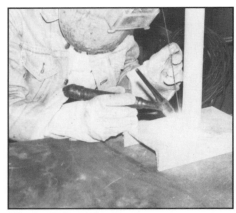

Weldor TIG-welds aluminum tube to thick aluminum channel section. TIG-welding aluminum is tricky because molten puddle doesn't change color; it gets shiny instead.

Not all aluminum is weldable. Airplane nose wheel was TIG welded (arrow). When wheel and tire was assembled, weld broke as tire was being inflated, below. Airplane nose wheel also suffered arc burns (outlined arrow) to welding table because workpiece wasn't properly grounded.

Mechanical cleaning is done with a stainless-steel wire brush, sandpaper or abrasive pads. Afterward, wash off the dust with soapy water and rinse with clear water. Then, wipe down seam area with alcohol, MEK or acetone.

Brumos Porsche 962 has many TIG-welded aluminum and chrome-moly parts. As with an airplane, effect could be fatal if a weld breaks. Photo courtesy Daytona International Speedway.

GUIDELINES FOR TIG-WELDING ALUMINUM

Material Thickness (in.)	Current (amps)	Tungsten Diameter (in.)	Weld Rod Diameter (in.)	Argon Flow (cfh)
0.020	25	0.040	1/32	16
0.040	34	1/16	1/16	18
0.063	50	1/16	1/16	20
0.080	75	3/32	3/32	20
0.100	100	3/32	3/32	22
0.125	125	3/32	1/8	25
0.250	150	1/8	1/8	35

Chart gives approximate settings for welding aluminum. Adjust amps to suit the particular conditions.

To clean chemically, dip the part into a 5% sodium-hydroxide solution for about five minutes. Rinse with clear water and air-dry. You can also spot-clean the weld area using this method.

- To eliminate the need for cleaning newly fabricated aluminum parts, use the more-expensive paper-covered sheet stock. Adhesive-backed paper-covered sheet can be cut, formed and fitted with the covering in place. When you're ready to weld, simply peel back the covering from the weld seam to expose super-clean aluminum and you're ready to weld.
- If TIG-welding aluminum castings or forgings, V-groove the joint all the way through. Or V-groove the joint from both sides and weld on both sides, if possible.
- If welding aluminum plate that's more than 1/16-in. thick, V-groove the joint for better penetration.
- Don't attempt to weld 2024 aluminum or other non-weldable aluminum alloys. Read on for a handy test to check the weldability of unknown aluminum alloys. Weldable alloys include 1100, 5052, 6061 and all castings.

Machine Settings—Refer to the drawing of a typical TIG machine, page 95, for the following settings to TIG-weld aluminum:

Ron Fournier recommends spending a little more for adhesive-paper covering when buying aluminum sheet. It prevents scratches, blemishes and oxidation. Forming and trimming can be done with paper in place. For welding, peel back covering in weld-seam area. Afterward, re-cover area until project is complete. Photo by Ron Fournier.

Preheat aluminum cylinder heads to about 350F (177C) before welding. Intake pipes were welded to heads on turbocharged Corvair engine.

- Polarity selector switched to A-C.
- High-frequency unit set to **CONTINUOUS**—an absolute necessity.
- Argon flowmeter set to 20 cfh, or check flow chart.
- Water cooling switched **ON**—if so equipped.
- Contactor and amp switches on **REMOTE**—if equipped with foot pedal.
- Set coarse amp adjustment to about 60 amps; fine amp adjustment to 70%. Reset amp setting(s) if heat is not right.

Also, you'll need to use the following items:
- *Pure* tungsten rather than 2%-thoriated tungsten for TIG-welding aluminum. As you may recall, thorium contaminates aluminum welds.
- Aluminum welding rod 4043.

Preheating—Not too many years ago, I toured one of the largest aircraft-engine repair shops in the world. I was shown the welding shop where cracked and broken cast-aluminum cylinder heads were repaired. They used a natural-gas oven to preheat three or four cylinder heads at a time before welding. The cylinder head was heated evenly to 350F (177C)—normal operating temperature. Heating above this temperature would weaken the metal.

Any aluminum more than 1/4-in. thick benefits from preheating before welding. A simple way to preheat is in a gas or electric oven with thermostatic control, such as a kitchen oven. Just make sure the aluminum you're welding is absolutely clean, or you'll stink up the house for a week!

It's usually not necessary to re-preheat during welding, because the weld heat keeps the part hot enough.

Weldability—Several aluminum alloys are not weldable, most notably 2024. Usually found in sheet form, this alloy is commonly used on airplane-wing and fuselage skins, wing ribs and fuselage bulkheads. If you're not sure about the alloy, look for previous welds. If it was welded before, it can be welded again. But, if there is no sign of welding, try a sample weld to test it. Finding a part to do sample welds on might not be easy, but I would even go to the trouble to find an unusable part to practice on to avoid ruining a repairable part. Race-car shops and airplane shops usually have scrap parts in a corner somewhere.

So, if an aluminum part is not weldable, repair it with rivets or nuts and bolts, or replace it.

Grounding Aluminum—When TIG-welding aluminum on a grounded table top, it usually arcs between the table and the work. Consequently, you may make a beautiful weld on a part that took all day to cut, shape and fit, only to turn it over and find arc burns and craters where it contacted the table.

Arcing can be prevented by grounding the workpiece directly or providing a simple ground for the work. I usually lay a heavy mechanical finger on the part to help hold it against the weld table. If you use a separate ground cable, make sure it's welding-cable size. A small-diameter cable will overheat from high amperage.

Weld Craters—When running an aluminum weld bead, don't break the arc or rapidly shut off the arc with the foot pedal as you reach the end of the seam. This will cause a small depression or crater at the end of the bead. Instead, back off the foot pedal slowly, then *lead*—move—the arc back to the already solidified bead. This will "freeze" the puddle while it's still convex.

Fitting Parts—Fitting parts close-

ALUMINUM ALLOYS

Alloy Number	Tensile Strength (psi)	Heat Treatable	Weldable
1100	15,000	No	Yes
3003	26,000	No	Yes
3105	23,000	Yes	Yes
5005	26,000	No	Yes
5052	41,000	No	Yes
5086	47,000	No	Yes
2024	61,000	Yes	No
6061	42,000	Yes	Yes
7075	65,000—75,000	Yes	No

Note: If welding becomes necessary after part is heat-treated, heat-affected area is likely annealed or softened. If heat-affected area is small, it usually does not require heat-treating again. But if affected area is large, part should be heat-treated again.

Refer to chart to determine if aluminum alloy is weldable.

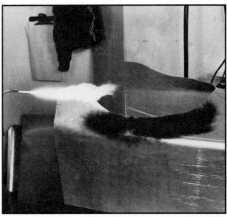

Area to be annealed is first covered with carbon from acetylene-rich flame. Work-hardened area is then annealed by playing neutral flame over area until carbon gradually burns off. When carbon is gone, aluminum is annealed. Metalworking can resume. Photo by Don Taylor.

ly is *crucial* when welding aluminum. Refer to Chapter 4, page 26, for details. You can fill a gap, but the back side of the weld will look like sand or even gravel because the molten aluminum solidifies too quickly and forms a lumpy bead.

Back-Gas Purge—TIG-welding aluminum does *not* require a back gas to purge oxygen and hydrogen from the weld as with stainless steel and titanium. Although it can improve the appearance of the back side of the weld bead, it will not noticeably improve weld integrity. Aluminum does not pick up atmospheric contamination as does stainless steel or titanium. And, aluminum will solidify in open air with nothing more than oxidation of the surface. Consequently, using back gas to weld aluminum is a waste of valuable time, money and equipment.

Heat-Treating After Welding—Aluminum is very easy to shape, form and TIG-weld in the dead-soft—O—condition. After these operations are completed, it is then heat-treated to give the soft aluminum strength and rigidity.

The 6061-O alloy is commonly used to make fuel-tank bulkheads, wing ribs and other parts for airplanes and race cars. The complete assembly or subassembly is then heat-treated in an oven or *brine*—salt—solution. In the oven process, 6061-O is heated to 950F (510C) for about 15—30 minutes and then air-cooled.

The brine-solution process involves heating the solution to 1000F (538C), a temperature at which it does *not* boil. The 6061-O alloy is immersed in the brine for 15—30 minutes. It is then immediately *quenched*—cooled—in 70F (21C) water. At this point, the aluminum isn't completely hard, but after sitting 24 hours at room temperature, it *age hardens* to full strength and hardness. Age hardening is also called *precipitation hardening*.

Note: After 6061-O is heat-treated, it is transformed into either 6061-T4 (solution heat-treated) or 6061-T6 (oven heat-treated).

The chart on page 103 details heat-treatment nomenclature. For example, for aluminum-alloy 6061-TX, the T means temper, the following number indicates heat-treatment type.

Annealing—Annealing aluminum is the process of heating it to 750F (399C) and allowing it to air-cool slowly to remove the effects of previous heat-treatment, so it can be cold-formed without cracking. It can be reheated after annealing.

TIG-WELDING MAGNESIUM

CAUTION: Magnesium can burn and support its own combustion. Water or dry-powder fire extinguishers will not put out a magnesium fire. In practical terms, the only way a magnesium fire can be extinguished is to wait for all the magnesium to be consumed. I once saw a race car with a magnesium transmission case and wheels burn to the ground. All that remained of the transmission were the axle shafts, gears, nuts and bolts. The magnesium parts burned to ashes. The intense heat melted the aluminum engine, leaving crankshaft and connecting rods laying in the dirt. A fully equipped fire truck could not extinguish the fire.

So, when welding magnesium, try to do so outside, away from flammables. Magnesium isn't likely to catch fire unless there are magnesium filings nearby—such as those created by machining or dressing the weld seam. If magnesium does catch fire, stand back and let it burn. You probably can't stop it.

Clean Before Welding—As with other metals, magnesium should be cleaned of all scale and corrosion in the weld-seam area before TIG-welding. Use some aluminum wool, steel wool or a stainless-steel brush to remove the white-powderlike corrosion.

If the corrosion can't be removed by mechanical means, use chemicals. Mix 24-oz chromic acid, 5-1/3-oz ferric nitrate and 1/16-oz potassium fluoride in 70—90F (21—32C) water to make a 1-gal chemical cleaning solution. Dip the part in this solution for three minutes, then remove it and rinse in hot water. Let the part air-dry before welding. Don't blow it off with compressed air. Compressed air may be contaminated with dirt, water and oil.

Stress-Relieving Magnesium— Magnesium alloyed with aluminum is susceptible to a unique phenomenon called *stress-corrosion cracking*. For example, if you drill a hole in magnesium and put in a tight-fitting bolt, the area corrodes, then cracks as a result of the corrosion. Check with the manufacturer to determine the alloy content of a particular magnesium alloy. Otherwise, you may need to have the magnesium analyzed metallurgically—an expensive process. These alloys must be heat-treated to remove the welding stresses, which would otherwise result in corrosion and cracking. See the chart at right for stress-relieving by heat-treatment.

ALUMINUM HEAT-TREATMENT NOMENCLATURE

TO	Completely soft, no temper	T6	Solution bath heat-treated, then artificially aged
T2	Annealed by heat to soften (cast only)	T7	Solution heat-treated, then stabilized
T3	Solution heat-treated, then cold-worked	T8	Solution treated, cold-worked, artificially aged
T4	Solution (salt bath) heat-treated		
T5	Artificially aged	T9	Solution treated, aged, then cold-worked

GUIDELINES FOR TIG-WELDING MAGNESIUM

Material Thickness (in.)	Current (amps)	Tungsten Diameter (in.)	Weld Rod Diameter (in.)	Argon Flow (cfh)
0.040	35	1/16	1/16	12
0.063	50	1/16	1/16	12
0.080	75	3/32	3/32	12
0.100	100	3/32	3/32	15
0.125	125	1/8	3/32	15
0.250	175	1/8	1/8	20

Listed are approximate values for welding magnesium. Make adjustments for conditions.

Follow above recommendations for setting up to weld magnesium.

STRESS-RELIEVING MAGNESIUM THROUGH HEAT-TREATMENT

MAGNESIUM SHEET			MAGNESIUM CASTINGS		
Alloy	Temp (F)	Time (min)	Alloy	Temp (F)	Time (min)
AZ31B-0	500	15	AM100A	500	60
AZ31B-H24	300	60	AZ63A	500	60
HK31A-H24	600	30	AZ81A	500	60
HM21A-T8	700	30	AZ91C	500	60
HM21A-T81	750	30	AZ92A	500	60
ZE10A-0	450	30			
ZE10A-H24	275	60			

To prevent cracking, magnesium sheet and castings must be stress-relieved by heating to given temperature and held there for times given. Heating is best done in an oven.

Plasma-arc cutter (right) is rapidly gaining popularity in auto-body shops because of its speed, accuracy and minimal warpage. Companion MIG welder is at left. Photo courtesy Linde Welding Products.

Miller Zip Cut cuts up to 3/8-in.-thick steel. Chart on control panel gives recommended cutting speed and machine settings for different metal thicknesses. Photo by Tom Monroe.

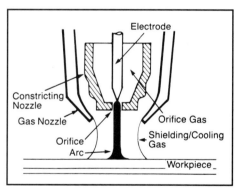

Unlike TIG welder, arc from plasma torch occurs in superheated gas column between tungsten tip and workpiece. Like TIG welding, shielding and cooling is done by flow of gas exiting torch between cup and constricting nozzle. Drawing courtesy Linde Welding Products.

Plasma is a gas heated to an extremely high temperature and ionized so it becomes electrically conductive. The plasma-arc welding (PAW) process uses such a gas to transfer an electric arc to the workpiece and to *constrict,* or contain, the arc for welding.

The Process—In the basic plasma-arc process, invented and developed by Linde, an electrode is located within a torch nozzle. This nozzle has an arc-constricting orifice. Inert gas, usually argon or nitrogen, is fed through the nozzle, where it is heated as high as 50,000F (27,760C), the plasma temperature range. The plasma arc emerging from the orifice is hot enough to melt *any* metal. In welding operations, a shielding gas is simultaneously introduced through a separate concentric passage of the torch. This protects the weld puddle from atmospheric contamination in a manner similar to a TIG-welding torch.

Because of the modest amount of skill required, plasma-arc is widely accepted for welding and cutting of ferrous and non-ferrous metals. The plasma arc's straight, narrow, column-like shape and high-current density mean that it is not critical to maintain a certain nozzle-to-workpiece distance to obtain weld or cut consistency. Also, greater nozzle-to-workpiece distance possible with the plasma techniques means better visibility of the workpiece for controlling the puddle or cut.

PLASMA-ARC WELDING

In plasma-arc welding, the plasma produces a much longer, hotter and easier-to-handle arc

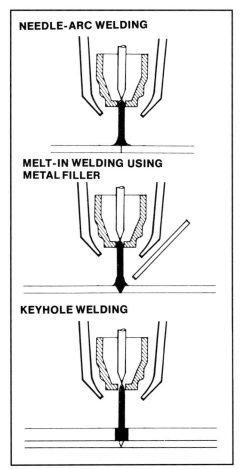

NEEDLE-ARC WELDING

MELT-IN WELDING USING METAL FILLER

KEYHOLE WELDING

Plasma-arc-welding processes, from top to bottom: needle-arc welding, melt-in welding using filler metal, and keyhole welding. Drawings courtesy Linde Welding Products.

than does a TIG welder. At low currents—under 100 amps—so-called *needle-arc welding* can be done. This long, needle-shaped arc is used to join very thin metal—0.001—0.125-in. thick. Results are comparable to those of other mechanized fusion-welding processes that require sophisticated controls to maintain precise torch-to-workpiece distance.

Higher currents can also be used in plasma-arc welding. Although a wider arc is generated, high-quality welds can be made on workpieces up to 1-in. (25mm) thick using currents up to 400 amps.

There are two modes of penetration in plasma-arc welding: *melt-in* and *keyhole.*

Melt-in utilizes the plasma arc for conventional, manual and mechanized fusion welding. The major advantages over TIG welding are better operator control of torch-to-work distance and the elimination of tungsten electrode contamination, because the electrode is protected inside the nozzle. High-quality, narrow butt welds or lap welds on joints up to 1/8-in. thick can be accomplished. Filler metal can be used.

Keyhole plasma-arc welding gives a long, narrow arc that completely penetrates the workpiece to form a keyhole at the center of the weld puddle. If a close-fitting butt-type weld seam is used, filler metal is not required. As the torch travels forward, molten metal forms at the leading side of the arc, flows around the arc and rises to form a small weld bead behind it. A complete weld on both the top and bottom surfaces is formed in one pass to give a 100% weld. The complete penetration of the workpiece thickness and movement of the molten metal purges impurities and gases from the weld prior to solidification. This gives the highest possible weld quality. Keyhole welding can be done with metals up to 1/4-in. thick.

Equipment—The equipment used for plasma welding is similar to that for TIG welding. At first glance, the torch looks the same. But concealed in the plasma-welding torch is a non-consumable tungsten electrode and a ceramic shielding-gas nozzle.

A power supply initiates the arc and contains the gas supply and water-cooling system. Capacity ratings range from 100 to 400 amps at 60 to 100% duty cycle. Direct current, straight polarity is usually used. Reverse polarity—positive electrode—is used with a water-cooled copper electrode. Although filler rod is usually added the same as in TIG welding, mechanized wire-feed systems can also be used for industrial applications.

The main difference between TIG and plasma-welding equipment is the gas supply. Although the same gas is used, two gas supplies are required with plasma welding, one for the orifice gas—plasma gas—and one for the shielding gas.

Which gas should be used depends largely on the type of welding technique—melt-in or keyhole—and the type of metal to be welded. For example, argon is used for welding steel—carbon, low-alloy or stainless—and aluminum with keyhole or melt-in techniques. A 75%-helium/25%-argon mixture is used if the material is over 1.8-in. thick when using the melt-in technique. For *reactive* metals such as titanium, use argon if the material is less than 1/4-in. thick. In other applications, use an argon/helium mixture; 50—75% helium for keyhole and 75% helium for melt-in.

Using Plasma Arc—Because the electrode doesn't extend from the nozzle, you can't touch the electrode to the work to start an arc as is possible with a TIG-welding torch. So, an arc must be initiated at the torch, either mechanically or electrically. The mechanical method involves extending the electrode until it touches the nozzle. The electrical method is done with a high-frequency, a-c power supply such as that used with TIG welding, or with high voltage superimposed on the welding current. Whichever method is used, the orifice gas is ionized, causing it to conduct the *pilot* arc current, initiating the torch-to-work arc. Once started, plasma-

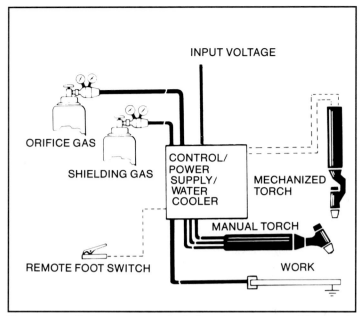

Two gas cylinders are required for plasma-arc welding and cutting. One cylinder contains normal shielding (secondary) gas, as in TIG welding. Other cylinder supplies restricted orifice gas. Drawing courtesy Linde Welding Products.

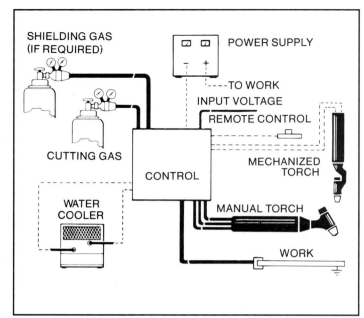

Plasma-arc-cutting setup: Water cooling and shielding gas are optional. Drawing courtesy Linde Welding Products.

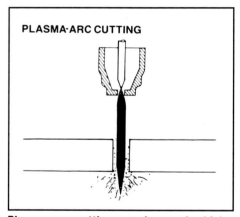

Plasma-arc cutting requires only high-energy gas column for cutting. Shielding gas is not required. Constricted-gas jet produces narrow, smooth kerf. Drawing courtesy Linde Welding Products.

arc welding technique is virtually the same as in TIG welding.

PLASMA-ARC CUTTING

Plasma-arc cutting uses a highly constricted, high-velocity arc that penetrates the metal similar to keyhole plasma-arc welding. However, up to 50,000 volts is used to melt the metal. Either compressed air from a shop air compressor or a blended, inert shielding gas is used to blow the molten metal out of the kerf. Because it uses a narrow, straight, column-like arc, there is minimal kerf width. And, because of the clean cut, the cut surfaces do not generally require cleanup. Metal up to 6-in. thick can be cut with a plasma-arc setup, depending on the type of metal and the arc current used.

Plasma arc can be set up to cut with nitrogen as its shielding gas. Although shielding gas is preferred for clean, no-oxidation cuts, shop air is also used because of its low cost. Because high-pressure, high-velocity shop air is extremely noisy, it can be distracting. The sound is similar to listening to a compressed-air blow gun at close range.

The plasma-arc cutting torch is excellent for auto-body shop work because it will cut through paint, undercoating, body putty and dirty metals. No precleaning is necessary. Because it doesn't use the oxidizing process to cut metal, plasma arc is ideal for cutting high-strength, low-alloy steel used in new unibody cars. This non-oxidizing feature also makes it a natural for cutting stainless steel, and non-ferrous metals such as aluminum, copper and brass.

The big advantage of plasma-arc

Ideal for auto-body work, plasma arc cuts through paint, rust and even undercoating with minimum warpage. Photo courtesy Linde Welding Products.

cutting is speed—up to 20 times faster than oxyacetylene. Available now are portable units complete with gas, torch and power supply, ready to plug in to 220-volt, 60-cycle, single-phase power.

MIG Welding

11

Welding race-car frame and roll cage is where MIG welder "shows its stuff." Although not as pretty as a TIG weld, MIG welding is much faster and as strong. Photo by Tom Monroe.

Metal inert-gas (MIG) welding is so called because of the types of electrode and shielding used. Unlike TIG welding, a MIG welder uses a consumable metal electrode. This electrode is a continuously fed wire that exits from the center of the welding torch, where a TIG-welder tungsten would normally be; thus, the name *wire-feed*. Typical wire sizes are 0.024, 0.030, 0.035 and 0.045 in. Up to 1/16-in.-diameter can be used with special equipment. CO_2 shielding gas is used in place of argon because CO_2 is less expensive. Hollow flux-core electrodes are used frequently, eliminating the need for gas shielding.

Fast & Clean—The major advantage of wire-feed welding is that it's fast. Unlike arc welding or TIG welding, you rarely have to stop to for a new welding rod. Also, its weld rate—inches per hour of weld bead—is fast, especially when compared to TIG. Another advantage of the MIG welder is clean welds—much cleaner than those possible from an arc welder

There are many MIG welders to choose from. Determine your needs, then go shopping. Linde's 225, 160 and 100 models have 225, 160 and 100-amp capacities, respectively. The 100 and 160 models are recommended for light-duty work such as that in auto-body shops. Model 225, suitable for general fabrication welding, can use flux-cored wire. Photo courtesy Linde Welding Products.

MIG welding is rapidly replacing gas welding in auto-body shops. It's faster, cleaner and heat-affected area is smaller. Concentrated heat makes MIG welding more suitable for welding high-strength low-alloy (HSLA) steel used in most late-model cars. Photo courtesy Linde Welding Products.

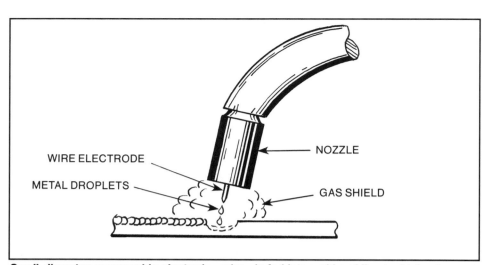

Small-diameter, consumable electrode—wire—is fed into weld puddle at a high rate, or up to 700 in. per minute (ipm). Instead of argon shielding gas typically used with TIG welding, carbon dioxide (CO_2) is preferred for MIG welding. Shielding gas is not required when flux-cored wire is used.

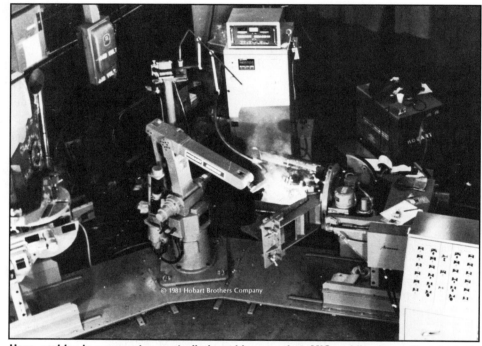

Here, *weldor* is a computer-controlled machine, or robot. MIG welding is readily adaptable to high production because of its continuous welding capability. Cabinet at right is power supply. Torch is at end of robot arm. Photo courtesy Hobart Brothers Company.

and its coated electrode. Because there's no slag, there's no dirt from it, nor is time used to remove it—another timesaver. These reasons make the MIG welder preferred by most production welding shops. So, if you want to work fast and easy, wire-feed is the best way to weld—but not for everything. It has its limitations—as I explain later in this chapter.

I once thought I could improve the production rate of airplane seat frames by wire-feed welding rather than using TIG. Unlike TIG welding where you must stop every 2—3 minutes to get another piece of filler rod, you just point the gun at the weld seam, pull the trigger, and the filler rod, arc and gas come out for hours on end without stopping.

To complete the airplane-seat story: I asked my welding-supply dealer to set up a wire-feed machine for welding the lightweight 4130 square-tube seat frames. We were ready for a test about a week

later. The machine worked perfectly. It was set up with just the right combination of a small, but accurately controlled power supply, wire-feed control and the smallest, lightest welding gun available. Although the welds weren't as pretty, we instantly reduced our production time by 75%! We could weld four times faster!

WIRE-FEED OPERATION

Here's how a wire-feed welder

operates: The gun is positioned over the weld seam at the same angle you would hold an arc-welding rod. Cup-to-work distance should be about equal to the distance across the cup opening. The gun trigger is then pulled, activating the d-c-current, positively-charged wire electrode—reverse polarity—and gas flow. (Straight polarity is rarely used because the arc is unstable and erratic. Also, penetration is lower.) The wire is simultaneously fed through the gun nozzle and contacts the grounded base metal, causing a *short circuit* and resulting arc to start. Resistance heating melts both the base metal and the end of the electrode. The wire then melts back faster than it is being fed to the base metal, momentarily breaking the arc and depositing metal. The arc force flattens the molten metal.

But the wire electrode is still advancing into the puddle, repeatedly arcing and melting off again and again. This on-and-off process occurs about 60 times per second, causing the characteristic buzz you hear. Some people describe a properly adjusted MIG welder as sounding like frying bacon.

WIRE-FEED MACHINES

The MIG welder is a simple, compact welding machine. It consists of the welding gun, power supply, wire-drive mechanism and control unit, shielding-gas supply and, for some heavy-duty units, water-cooling system.

Gun—In place of the TIG torch or arc-welder stinger is a *gun*. The typical gun looks like a pistol and directs the filler metal and shielding gas to the weld seam. Service lines running to the gun include an electric-power cable, electrode conduit, and gas hose, if used. Heavy-duty industrial-type guns also have water lines for water-cooling. Otherwise, gun cooling is done with air. Electric power is transferred to the wire electrode via a sliding contact with the copper electrode-guide tube in the gun.

The gun nozzle, which is usually interchangeable, determines the gas-shield coverage of the weld puddle. Nozzle-orifice size varies from about 3/8 to 7/8 in. (10 to 22mm). A larger orifice gives additional shielding, as does a larger TIG-torch cup.

Power Supply—Almost all wire-feed welders supply d-c current. This requires a transformer-rectifier when using an a-c power source. Depending on the machine, output can range from 15 to 1200 amps. Required power supplies typically range from 110 to 200/230 volts, or all the way up to 575 volts, depending on machine output. Duty cycle is either 60% or 100%.

Wire-Drive Mechanism & Control Unit—The wire-drive mechanism is relatively simple. It consists of a wire spool and d-c-motor powered *drive rolls*—two wheels that run against each other with the wire in between. Sometimes, two sets of wheels are used. The drive-roll mechanism pulls the wire off the spool and pushes it through the conduit to the gun at a weldor-adjusted rate.

The MIG-welder control unit regulates arc starting and stopping, as well as wire-feed, gas-

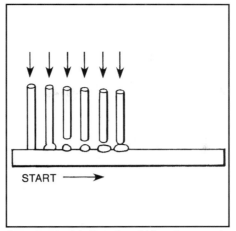

Arc starts at left and weld bead moves to right. Wire continuously advances and melts as it contacts base metal. With machine adjusted correctly, cycle occurs smoothly and quickly, sounding as if it were bacon frying.

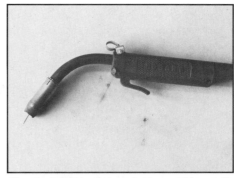

To start filler wire and gas, weldor pulls trigger. Note wire extending from tip. Because wire must be cleanly cut to start welding, always have wire cutters close at hand. Always cut end of wire when stopping and restarting weld. Photo by Tom Monroe.

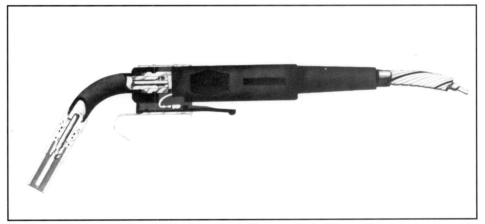

Air-cooled MIG gun: Wire electrode and shielding gas is fed to gun through power cable. Photo courtesy Linde Welding Products.

Lightweight, self-contained Miller 200 is on casters so it easily can be moved to the work. Gas cylinder is mounted at rear of welder. Wire-feed unit with spool is mounted on power supply.

Linde power supply and wire-feed control is typical of most 200/230-volt machines. Note argon flowmeter on wall at left.

Wire spool and drive mechanism using single set of rolls: Weld-Aid Product's applicator (arrow) applies lubricant to wire immediately ahead of drive rolls to prevent binding and reduce friction. Lubricant should not be used on aluminum wire. Photo by Tom Monroe.

CO_2 pressure regulator/flowmeter monitors cylinder pressure and controls and monitors gas flow to torch. Photo by Tom Monroe.

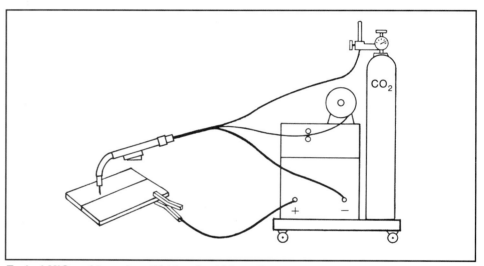

Typical MIG-welder setup: Welder should be self-contained for portability. A 50—100-ft. power cord allows welder to be used outside.

flow and, sometimes, water-flow rates. It also synchronizes these functions. Usually, there's a *jogging* feature that feeds the wire to or through the gun while not welding.

Similar to the wire-jogging feature, the control unit also has a shielding-gas purge switch to manually control gas flow. In addition, timers control preweld and postweld gas flow automatically. The purge switch can override the automatic timers. Another timer controls water flow, if water cooling is used. Finally, a wire-feed brake stops

the electrode the instant the gun switch is released. This prevents wire from being fed to the puddle when the arc is interrupted.

Shielding Gas—Except for CO_2 used in place of argon, a MIG-welding shielding-gas setup is similar to that used for TIG welding. Not only is CO_2 less expensive than argon, it also has superior heat conductivity.

Regardless of the gas used, constant pressure and flow must be maintained while welding. Also, you must be able to adjust pressure and flow for different applications. Therefore, the gas cylinder must

be equipped just as if it were used for TIG welding, shown above. Although different gases use different flowmeters or flowgages, the CO_2 must have its pressure and flow regulated.

WIRE-FEED WELDER TYPES

What to Buy?—When considering a wire-feed welder, remember the old adage: You can't drive a railroad spike with a tack hammer or a tack with a sledge hammer. You must match the machine to the job. If you have both heavy- and light-duty welding to do, you need

Compact Cyclomatic 200SMF MIG welder has features of much larger machines. Unit can also be used for TIG or arc welding, or carbon-arc gouging. Photo courtesy Cyclomatic Industries Inc.

MIG-welder control panel is simple when compared to other types of welders. From top to bottom, wire-feed speed adjustment, on/off switch and voltage control. Gas flow must also be set. Photo by Tom Monroe.

AWS# E71T-GS FLUX-CORE WIRE-FEED CAPABILITIES

Steel Thickness (in.)	Amps	Volts	Wire-Feed Speed (in./min)	Wire Stickout (in.)
18 gage (0.048)	95	13	75	1/2−3/4
16 gage (0.060)	95	13	75	1/2−3/4
11 gage (0.120)	150	16	133	1/2−3/4
1/4 in. (0.250)	160	17	155	1/2−3/4

When using flux-core wire for welding steel, refer to chart to adjust machine. Chart courtesy Hobart Brothers.

two welding machines, one heavy and one light.

Unlike TIG, you cannot buy a power supply now and add on later. Wire-feed welders are self-contained, not made up of separate components. So, it's important to decide your MIG-welding needs up front.

110-Volt—There are some acceptable 110-volt wire-feed welders available. Hobart Brothers Company makes a light-duty portable wire-feed welder for sheet-metal and body-shop work that plugs into 110-volt service. This welder has some limitations, but it's effective at doing what it was designed for—welding 24-gage (0.0239-in.) sheet metal. It will not weld heavy-gage stock, such as that used for trailer hitches or farm equipment. Some of the specs for this machine are:

- 100-amp maximum.
- 0.024- and 0.030-in. steel wire.
- 0.035-in.-and-smaller aluminum wire.
- Gas-flow timer allows making *stitch welds*—tack welds about 1/2—1-in. long—and tack welds. Timer guarantees faster gas shutoff after each weld is completed.
- Wire-feed speed control regulates arc heat.
- Unit is self-contained except for CO_2 shielding-gas cylinder. CO_2

is used instead of argon because it's effective and inexpensive. Where high-quality welds are required in 4130 steel, a mixture of 75% CO_2 and 25% argon may be used. This comes premixed in one cylinder from welding-supply stores.

200/230-Volt—Several companies make conventional 200/230-volt wire-feed welders. These are the most common machines. Most machines this size can be made portable, but rarely will they plug into a wall socket. Chances are, you'll have to change the socket to fit the plug on the machine. They can handle 0.024-, 0.030-, 0.035- and 0.045-in. wire.

No CO_2 Gas Required, Your Choice—If you use flux-cored wire—0.045-in. diameter—no gas-shielding is required. It's similar to stick welding, except you don't have to stop for new rod. Flux is on the inside of the electrode, so it doesn't chip and flake off.

When welding any kind of sheet steel, plain or galvanized, this wire gives good weld performance with little spatter. It can also weld thicker metals at higher volt and amp settings. Refer to the chart on this page for your specific welding application.

575-Volt Heavy-Duty Welder—The *minimum* these 800-amp machines will weld is 100 amps. They have a 100% *duty cycle* at 800 amps! You can operate the machine at 800 amps all day, without cool-down, if necessary. If you have some heavy-duty welding to do, this type of machine will handle the job. But, it's definitely not for the average home weldor! If you think you absolutely must have one, try it first. Remember, you don't need a sledge hammer to drive tacks. And you don't need a 100-amp-minimum welder to build a utility trailer.

WIRE-FEED WELDING STEEL & ALUMINUM

Most welding instructors give you about five minutes of discussion about a MIG-welding ma-

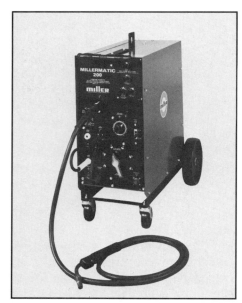

Miller 200/230-volt machine can be fitted with gun that pulls wire through cable. Such a gun is often used with soft wire—aluminum—or a long cable, which causes high friction. Photo courtesy Miller Electric Mfg. Co.

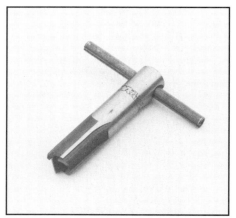

Weld-Aid's special reamer will quickly remove weld spatter from MIG nozzle. Photo by Tom Monroe.

Install lubricating pads to reduce friction and prevent binding, page 110. Use dry pads only to remove dust from aluminum wire. Photo by Tom Monroe.

To reduce weld-spatter buildup, coat MIG nozzle with spray or dip anti-spatter. Photo by Tom Monroe.

chine and then tell you to go run a few beads on a piece of steel or aluminum. And that will be the extent of the instruction. That's how easy it is to start wire-feed welding.

However, there's more to wire-feed welding than pointing the gun at the weld seam and pulling the trigger. Problems can occur if you don't take precautions. The wire may get tangled in the drive mechanism or stick in the collet. The cup on the gun may fill with spatter, making weld quality erratic. Just call it Murphy's Law of MIG welding.

Adjustments—There are also several adjustments that have to be exactly right or the weld will end up looking worse than if it were done with an E-6011 arc-welding rod in the hands of a beginner. Here are some adjustments you'll have to make:

- Wire-feed speed.
- Power—amps.
- CO_2 or argon gas-flow rates.
- Wire size.
- Amps and volts *while welding*.

You'll need a helper to monitor these readings as you weld, and make adjustments as necessary.

Maintenance—Lack of maintenance causes frequent wire-feed-welding problems. The cup *will* get dirty—it *must* be kept clean. The cup and nozzle must be cleaned of spatter regularly. Special sprays and jellies are made for this purpose. To keep your MIG welder operating trouble-free, perform these simple maintenance operations:

- Keep the cup clean. Normal weld spatter will clog the cup quickly, blocking gas flow and therefore cause the welder to be unprotected from air. To prevent this, you should coat the inside of the cup with an anti-spatter spray or gel such as Nozzle-Kleen or Nozzle-Dip Gel from Weld-Aid Products. As a less-expensive alternative, many weldors use Pam, an aerosol cooking-oil substitute available at supermarkets. If the nozzle does get dirty, clean it. Special ream-like MIG-welder nozzle cleaners are available.
- Keep the drive gears and rollers clean. Copper-plated wire will clog the drive rolls in a hurry, so avoid copper-plated wire, if at all possible. Every 2—3 hours of welding time, check the drive rollers for metal filings, and brush them away before continuing. To reduce wire drag and clogging, lubricate the wire. Use a treated-felt applicator to lube steel wire; untreated applicator for aluminum wire.

PREPARE TO WELD

Before you start welding, attach a small notepad or a sheet of paper to the MIG machine to record settings that work for you. Copy the following chart to make your notes.

Settings vary from one machine to another. Even though two machines are the same model, they are not the same machine. Some examples of these variables are:

- How fast the wire filler rod is fed at a setting of 1, 2 or 3.
- How hot the wire gets when the volt and amp settings are adjusted to mid-range.
- How much shielding gas is supplied to the weld puddle when adjusted for 20 cfh.

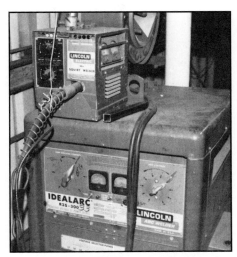

Lincoln's *Squirt Welder* wire-feed mechanism used to convert d-c arc welder for MIG welding.

MIG nozzle is in bad shape. Nozzle should be replaced to obtain better welds.

MIG welding produces a lot of smoke and sparks. Always make sure area is well-ventilated and free of flammables. Weldor should wear protective helmet, leather gloves and long-sleeve shirt, buttoned to the neck. Photo by Ron Fournier.

WIRE-FEED ADJUSTMENT					
Metal Thickness (in.)	Amps	Volts	Gas Flow	Wire Size	Wire Speed
0.020	_____	_____	_____	_____	_____
0.030	_____	_____	_____	_____	_____
0.040	_____	_____	_____	_____	_____

Make copies of chart for recording MIG-welder setups. You'll then be able to make quick setups when doing similar jobs.

Not only will one machine vary slightly from another, each project will vary in adjustment requirements. For example, if I were describing how to drive my car, I couldn't tell you how far to depress the accelerator to maintain a certain speed. It's the same with MIG welders. Although you can come close on your initial settings, trial and error are required to get them exactly right. So pick a reasonable starting setup such as described in the chart for flux-core wire on page 111. Fine-tune the machine after making a few test beads. Practice on some scrap-metal pieces of the same metal you're welding until you can determine the right settings.

As you find the correct settings for each welding situation, record them on your "Wire-Feed Adjustment" chart. Then, next time you're welding 0.020-in. steel or whatever, you can set up your machine perfectly by just referring to your chart.

Adjust by Sound—I adjust my welder by listening to the arc. Seriously, when it's right, the arc sounds almost like bacon frying on a grille over an open fire. Of course, you should also check for proper weld deposit and penetration. Once you're satisfied with the arc sound and weld quality, have someone read the volt and amp gages *while you are welding*. Record these numbers in your chart, too.

Wire Cutters—Always keep a pair of small diagonal cutters handy. You'll soon learn why. When you pull the trigger to start the arc and it doesn't, you'll end up with a "pile" of wire. Even though the arc doesn't start, shielding-gas

SPECIAL NOTE ON HIGH-STRENGTH STEEL CAR BODIES

To reduce weight for improved fuel economy, many late-model cars use high-strength steel in the body structure. Because the high-strength steel has excellent tensile strength—about 40,000—120,000 psi compared to mild steel's 30,000 psi—panel thickness can be reduced, resulting in considerable weight savings.

But there's a hitch. High-strength steel cannot be gas-welded or brazed. It will harden and crack. Because of its unique grain structure, it must be arc-welded with a low-hydrogen stick electrode such as E-7014, or wire-feed welded. Ford Motor Company and Chrysler Corporation, for example, recommend MIG welding. Here are some of the additional reasons they give for recommending the wire-feed process:

- Welds are made quickly on all types of steel.
- Low current can be used, resulting in less distortion of sheet metal.
- No extensive training is necessary.
- Wire-feed equipment is no bulkier than a set of oxyacetylene cylinders.
- MIG spot welding is more tolerant of gaps and misfits in seams.
- Severe gaps can be spot-welded by making several spots atop each other.
- Simple to weld vertically and overhead.
- Metals of different thicknesses can be welded easily with the same-diameter wire.
- Almost all auto-body sheet metal can be welded with one wire type.

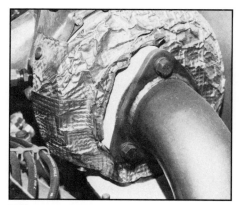

MIG-welded turbocharger exhaust flange of my Corvair-powered Cessna airplane shows how a good MIG weld should look.

WIRE-FEED STAINLESS STEEL AISI Series 200 and 300 stainless				
Plate thickness (in.)	1/8	1/4	3/8	1/2
Wire Size (in.)	1/16 (.060)	1/16 (.060)	1/16 (.060)	3/32
Current DCRP*	225	275	300	325
Wire-feed Speed (ipm)	140	175	200	225
Arc Speed (ipm)	19-21	19-21	15-17	15-17
*D-c current, reverse polarity				

flow and wire feed do. You can't release the trigger fast enough to prevent this, only minimize it. Before you can continue welding, you must cut off the excess wire. Simply snip it off with cutters and continue welding.

MIG-WELDING ALUMINUM

The best new racing yachts are made of 1/8—3/16-in. 5083 aluminum plate. The aluminum plates are joined by welding with wire-feed welders. The electrode wire is 3/64-in. ER-5356. Argon shielding and reverse polarity, d-c current are used. So, you can weld aluminum using your wire-feed welder.

If you want to weld aluminum, here are some specs you can start with:

- Base metal   1/8—3/16-in. 5083 aluminum plate
- Welding wire   ER-5356, 3/64-in. diameter
- Shielding gas   Argon
- Polarity   D-c reverse

WIRE RECOMMENDED FOR STEEL

The wire diameter should be 0.030—0.035 in. for most automotive sheet-metal work:
- Airco A-681
- American Chain and Cable AS-28 or AS-18
- Chemetron SPOOLARC-88
- Hobart Brothers HB-28
- National Standard NS-115
- Reid Avery 70S-G
- Linde-86

NOTE: Most high-strength steel used in late-model cars is confined to body structures, reinforcements, gussets, brackets and supports. In most cases, the outer panels remain regular mild steel and can still be gas-welded or brazed.

MIG-WELDING STAINLESS STEEL

The correct wire for MIG-welding stainless steel depends on its alloy. In most cases, 300-series MIG-welding wire will work with the more-common 300-series stainless steel. If the alloy is unknown, try ER-308—a general-purpose stainless wire. You can even weld mild steel with ER-308 wire. The weld on mild steel will be much less ductile than the base metal. Some common stainless-steel wire numbers are ER-308, ER-309, ER-310, ER-312, ER-348, ER-410, ER-420, ER-430 and ER-502.

Each number represents different carbon and alloy contents. For instance, the chemical content of ER-308 is carbon 0.08%, chromium 20%, nickel 10%, manganese 2%, silicon 0.50%, phosphorus 0.03%, sulfur 0.03%.

12

Welding Rods & Fluxes

Choosing the correct filler rod for welding a specific material is as important as knowing how to weld. You can buy scores of different types of welding rods at your local welding-supply outlet. There are hundreds of different types. So, it's obvious that a weldor needs some help in choosing the correct welding rod.

First, find out what kind(s) of metal you are welding. Refer to Chapter 1. Then, decide which welding process is best for that metal. In many cases, you have several choices.

For instance, you may want to weld or braze a broken office chair. If the break occurred at other than a previously welded joint, you'd have at least four choices of how to fix it: gas-weld with oxyacetylene, braze with oxyacetylene, arc-weld with small-diameter welding rod, or TIG-weld it.

In the case of repairing a break at an old weld, you'd first have to determine how it was welded, and then probably reweld it using the same method. Why? Some welds are not compatible with others. To determine the previous metal-joining method, make a visual inspection with a magnifying glass. An arc weld usually has some slag and spatter; a gas weld has flaking. TIG- and wire-feed weld beads are clean. Brass braze is a gold or bronze color.

The easiest way to weld the broken chair, assuming it's a new break, would be to gas-weld it with either bare steel wire or copper-coated steel wire—if you have access to a gas welder. Any of the other three choices would work.

To choose the correct welding rod, first determine the type of welding equipment available, then pick the correct welding rod. Welding-rod choices are grouped according to the specific welding equipment.

In my home workshop, I have about ten different kinds of gas-welding and brazing rod, six kinds of arc-welding rod, and four or

Whether stored in plastic bag with desiccant, an old refrigerator with a 100-watt light bulb inside, or an industrial oven such as this, coated welding rod must be kept dry. Photo courtesy Phoenix Products Company, Inc.

WELDING PROCESSES/FILLER METALS

	Equipment	Process	Rod to Use
1.	Gas torch	Lead solder	Lead alloys
2.	Gas torch	Silver braze	Silver alloys
3.	Gas torch	Brass, aluminum and bronze brazing	Brass, aluminum and bronze
4.	Gas torch	Fusion welding	Steel, aluminum
5.	Arc welder	Fusion welding	Stick electrodes
6.	TIG, plasma-arc welder	Fusion welding	Bare steel, aluminum and titanium
7.	Wire-feed welder	Fusion welding	Steel-, aluminum- and titanium-wire spool
8.	Spot welder	Fusion welding	No rod used.

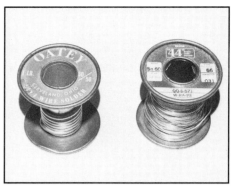

Use 50/50-alloy lead solder (left) with a separate flux. The 60/40 solder (right) has rosin core.

SOLDER - FOR FLAME SOLDERING WITH OXYACETYLENE			
Number	**Name**	**Use For**	**Method**
50/50	Lead solder	Solder copper tubing	Torch
10/90	Lead solder	Solder brass radiators	Torch
20/80	Lead solder	Fill seams in auto bodies	Torch
All State No. 55	Aluminum solder	Solder aluminum window frames, zinc carburetors	Torch
Welco 1509	Aluminum solder	Joins aluminum, die-cast stainless steel, copper & brass to themselves, each other	Torch

Use recommended solder or equivalent for soldering with gas torch.

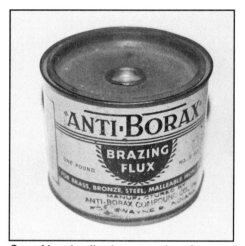

Can of brazing flux has served me for more than 10 years! I've used it for building several race cars and various repair projects. Keep container tightly sealed so moisture will not contaminate flux.

five kinds of solder. A commercial welding shop would have a similar assortment of filler materials. For TIG welding, you want at least ten different kinds and sizes of welding rod for welding common steel and aluminum alloys of various thicknesses.

In addition to being used as a filler material, welding rod has numerous other uses. For instance, you can use an arc welder and welding rod to build up a worn surface, such as a crankshaft journal.

Or, welding rod can be used to add a hardened surface.

LEAD SOLDERING WITH GAS WELDER

50/50 wire solder typically comes in spools and is used with brush-on flux to solder copper tubing with an oxyacetylene torch. Melting temperature is extremely low; 250—400F.

Wire solder is available in several different alloys: 10/90, 15/85, 20/80, 40/60 and 63/37. In the recommended example, 50/50 indicates 50% tin and 50% lead composition by weight. High lead-content solders such as 10/90, 15/85 and 20/80 are used for sealing brass auto radiators and filling seams in steel automobile bodies. These solders are used primarily with acid-base fluxes.

Inorganic fluxes are the strongest, yet most corrosive. Because inorganic fluxes are corrosive, they should be used only where it's easy to remove them after soldering. They are usually made of salts such as ammonium chloride or zinc chloride dissolved in water. Use inorganic fluxes only when soldering copper or steel buckets, or small pieces that are easy to dip or wash clean.

Medium-strength organic fluxes are used for soldering copper tubing and brass as well as steel. Usually, they are called *acid fluxes* because they are made from *glutamic* or *stearic* acid. Otay brand flux is called *No. 5 soldering paste,* and is for cleaning and fluxing all

metals except aluminum, magnesium and stainless steel.

Rosin-flux solders are the weakest type and *must not be used for flame soldering.* The base for rosin flux comes from pine-tree resin. It activates when heated and deactivates as it cools. *Use rosin flux only for soldering electrical and electronic components.*

ALUMINUM SOLDERING WITH GAS WELDER

Welco 1509 is a low-temperature (500F/260C) solder for joining aluminum, zinc, die-cast metal, copper, brass, stainless steel and other metals to each other, or to themselves. It requires a flux for cleaning and soldering. I use **Welco 380** flux. This solder has a *tensile strength*—stress at which it breaks while under tension—of 29,000 psi. It comes in 1-lb wire spools and is available in 1/16-in., 3/32-in. and 1/8-in. sizes. Welco 1509 can be used on aluminum car radiators, A/C evaporators and condensors, and other similar metals.

All State No. 55 Rubbon is a low-temperature self-fluxing solder alloy for aluminum window frames, zinc-based carburetors and outboard-motor housings. It comes in 1/8-in. diameter rods and melts at 700F (371C).

SILVER BRAZING WITH GAS WELDER

All State No. 101 and 101FC

Trucote braze are general-purpose silver-brazing alloy rods. They have a high tensile strength of 52,000 psi with a working temperature of 1145F (618C). Either rod may be used to join almost any metal with a melting temperature above 1150F (621C). No. 101 or 101FC Trucote braze rod is especially good for soldering copper, brass, steel, stainless steel or aluminum. These rods may be used bare with a separate flux or can be bought with a blue-colored flux coating. They come in diameters of 1/16, 3/32 and 1/8 in.

Welco 200 braze is a high-strength, 56% silver alloy for ferrous and non-ferrous metals. Its bonding temperature is 1155F (624C); tensile strength is 85,000 psi! By contrast, the tensile strength of chrome-moly steel is only 70,000 psi! This silver brazing rod can be used for race-car suspensions and airplane-wing ribs. It requires liquid flux.

Welco 200 flux for brazing comes in a 12-ounce plastic jar in paste form. It can be removed with warm water after brazing.

BRASS, BRONZE & ALUMINUM BRAZING WITH GAS WELDER

All State nickel/bronze braze rod has a high tensile strength of 85,000 psi. This rod melts at 1200—1750F (649—954C), so it's a little harder to work with than lower-temperature brazing rods. Regardless, braze rod does a good job when used properly, for assembling bicycle frames and ornamental railings. It should not be used on chrome-moly steel because it penetrates the grain of the base metal and cracks it.

All State No. 41FC braze is a high-quality rod with a tensile strength of 60,000 psi. It is flux-coated, eliminating the inconvenience of using separate flux. No. 41FC brazing rod is widely used to make bicycle frames, race-car frames and for repairing auto bodies. If exposed to the atmosphere over a period of time, the

| BARE & COATED BRASS & ALUMINUM BRAZING ROD FOR OXYACETYLENE ||||
Number	Name	Use For	Method
All State Nickel Bronze	Brazing rod 85,000 psi	Bicycle frames, race-car frames, general use	Torch
All State No. 41FC	Brazing rod 60,000 psi	Body shops, sheet metal, bicycle frames	Torch
All State No. 31	Aluminum brazing rod	For 1100, 5052, 6061 aluminum, fuel tanks, oil tanks, trailer bodies	Torch
All State No. 33	Aluminum brazing rod	For brazing aluminum castings, cylinder heads	Torch
Welco 10	Aluminum brazing rod	30,000-psi tensile strength for sheet aluminum	Torch

| BARE & COATED SILVER BRAZING ROD FOR OXYACETYLENE ||||
Number	Name	Use For	Method
All State No. 101	Silver brazing rod	Joins copper, brass, steel, stainless steel, aluminum	Torch
All State No. 101 FC	Flux-coated silver braze	Joins copper, brass, steel, stainless steel, aluminum	Torch
Welco 200	85,000-psi Silver braze	Carbide-tool tipping, airplane structural stainless steel	Torch

Refer to Chapter 7 for correct process to use with each brazing rod.

flux coating will flake off and the rod becomes unusable. Therefore, I buy flux-covered rod for jobs I expect to complete within a few days.

All State No. 31 aluminum braze is primarily meant for use with thin sheet aluminum such as fuel tanks, oil tanks and truck, race-car and trailer bodies. It can also be used to repair aluminum irrigation pipe. No. 31 braze rod should not be used on 2024- or 7075-series aluminum. It comes in 18-in. lengths, and 1/16-in., 3/32-in. and 1/8-in. diameters. It has a high melting temperature of 1075F (579C). A good flux such as *All State No. 31* should be used.

All State No. 33 aluminum braze is for brazing aluminum castings. This filler will repair cracked or broken castings and will fill holes or build-up areas that have been worn away or broken off. Use *All State No. 31* flux with this rod.

Welco 10 aluminum braze is a 30,000-psi tensile-strength alloy used on aluminum sheet. Use *Welco No. 10* flux.

Anti-Borax flux comes in 1-lb cans and is used for brass brazing of brass, bronze, steel and cast iron. It can be used in paste form by mixing with water. Or, it can be used in its dry-powder form by simply dipping the brazing rod into the can after heating the rod with the torch. I like to use the heat-the-rod-and-dip method for best results.

Glass jar is used for aluminum flux because this type of flux reacts with tin cans. Mix powdered flux with water to make a paste. Paint paste on base metal just before making weld puddle. Also, dip bare welding rod into paste as you weld.

BARE WELDING ROD FOR WELDING STEEL & ALUMINUM

Number	Name	Use For	Method
Welco 120	Aluminum welding rod	Torch welding (not brazing) aluminum 1100, 5052, 6061 gas tanks, oil tanks	Oxyhydrogen, Oxyacetylene Torch
No. 1100	Aluminum welding rod	Torch welding (not brazing) aluminum 1100, 5052, 6061 engine cowls, sheet alum.	Oxyhydrogen, Oxyacetylene Torch
Welco W-1060	Welding rod #7 steel	Copper-coated. Won't rust as fast if not used promptly as bare steel. Use on mufflers, exhaust.	Oxyacetylene Torch
—	Airco or Oxweld	Bare steel. Use for welding airplane-fuselage, race-car parts. General fusion welding.	Oxyacetylene Torch

Refer to Chapter 6 for correct process to use with each welding rod.

GAS-WELDING ALUMINUM OR STEEL

Welco 120 welding rod is a high-quality alloy for fabrication and repair of most weldable aluminum alloys. This welding rod is easy to control, and the bead solidifies rapidly, producing a nice-looking weld. Use oxyhydrogen if you plan to do much aluminum welding. Use *Welco No. 10* flux when welding with this rod.

No. 1100 Aluminum welding rod is the most common rod to use for all-purpose gas welding with a separate flux. Use *Oxweld Aluminum flux No. 725FOO.* This flux comes in 1/4-lb jars, and the label states "for all aluminum welding."

All State Sealcore is a unique tubular aluminum welding rod with the flux contained inside its hollow core. This welding rod is claimed to be the most versatile of all aluminum welding rod for torch welding. It comes in 1/8- and 3/16-in.-diameter rods, 20- and 32-in. long. It is supposed to be good for field work such as repair of irrigation pipe and farm equipment.

Coat Hangers: For many years, weldors have used coat hangers to oxyacetylene-weld car fenders or

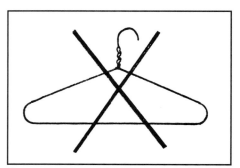

Never use coat-hanger wire for gas welding. Hangers are for hanging clothes; welding rod is for welding.

anything else they could think of. *Don't do it!* Paint on coat-hanger wire contaminates the weld. And the alloy in the wire is unknown. Usually, coat hangers are so brittle that they break when you try to straighten them. I suspect they are made of the cheapest steel available. You wouldn't want a weld to crack because of poor-quality filler material.

Welco W-1060 mild-steel rod is a good choice for gas-welding mild steel. This gas-welding rod is copper-coated. It is available in 1/16-, 5/64- and 1/8-in. diameters, 36-in. long. My last invoice from an Airco dealer called this *#7 Steel.* It works well for automobile exhaust systems and, if used carefully, OK for gas-welding non-structural 4130 or 4340 aviation steel. But the copper coating

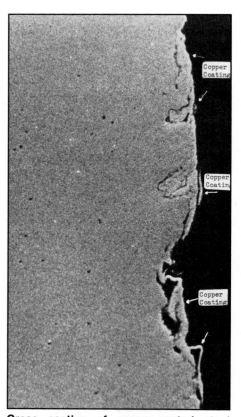

Cross section of copper-coated steel welding rod magnified 3000 times. Copper coating looks foreign to rest of metal. It can do the same thing to your weld. Although it's OK to use for welding exhaust systems and metal furniture, don't use copper-coated rod to gas-weld expensive parts. Photo courtesy United States Welding Corp.

contaminates the weld if the rod is not used carefully. See the above photo of a scanning electron micrograph of the copper coating.

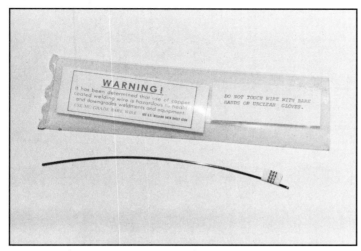

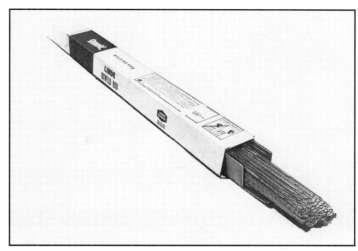

Package containing high-quality steel welding wire warns; DO NOT TOUCH WIRE WITH BARE HANDS OR UNCLEAN GLOVES. Welding rod must be kept clean and free of rust. Also note warning that copper-coated welding rod can be hazardous to your health.

Bare-steel rod for TIG welding is boxed to keep rod free of contamination, including dust! I use Oxweld rod for TIG- and gas-welding 4130 steel. Photo courtesy Linde Welding Products.

Airco or Oxweld bare mild-steel rod is what I use for torch welding, gas welding or oxyacetylene welding 4130 steel. And it's what the Experimental Aircraft Association recommends in their welding schools for welding airplane parts. It comes in 1/16-, 5/64- and 1/8-in.-diameter rods, 36-in. long. One pound that will last for many hours costs about as much as two soft drinks. Bare mild-steel rod is a real bargain, but will rust faster than copper-coated rod if not kept moisture-free.

ARC WELDING

The charts on pages 119 and 121 indicates how to identify arc-welding rod. Welding-rod diameter refers to the wire size, not the diameter of the coating. Wire diameter is measured easily at the holder-end of the rod.

E-6011 is the easiest-to-use all-purpose welding rod for arc-welding mild steel with a 220-volt a-c buzz-box arc welder in all positions. This is a good welding rod when the work is dirty or oily, and you don't have time to make the job pretty. The E-6011 rod produces a considerable amount of unsightly *spatter*—small globules of molten metal that stick to the base metal—in the area of the weld. I recommend E-6011 arc-welding rod for making repairs on farm equipment.

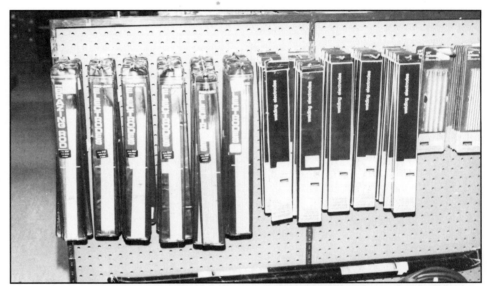

If you need welding rod on weekends, you may find packaged brazing rod and stick electrode in department stores.

STICK-ELECTRODE WELDING ROD FOR MILD STEEL

Coating Color	End Mark	Spot Color	AWS Rod Number	Use For Welding	Polarity + = Reverse − = Straight
Brick Red	—	—	E-6010 or 5P	Pipe, oil-field work	d-c welder + (only d-c)
White	—	Blue	E-6011	General purpose, spatters a lot	a-c or d-c d-c +
Dark Tan	—	Brown	E-6013	General purpose, clean work	a-c or d-c d-c −
Gray Brown	Black	Brown	E-7014	Sheet metal, general maint.	a-c or d-c d-c −
Gray	—	—	E-7018	All position, clean work	a-c or d-c d-c +

Refer to Chapter 8 for correct process to use with each welding rod.

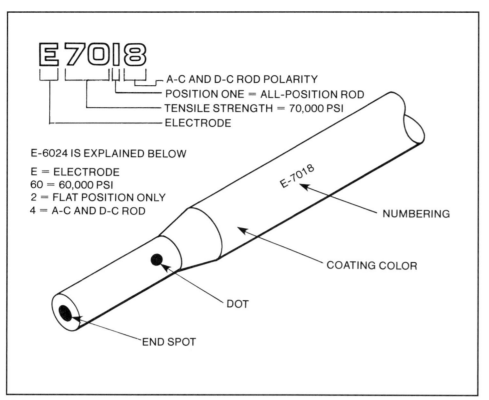

E7018

— A-C AND D-C ROD POLARITY
— POSITION ONE = ALL-POSITION ROD
— TENSILE STRENGTH = 70,000 PSI
— ELECTRODE

E-6024 IS EXPLAINED BELOW

E = ELECTRODE
60 = 60,000 PSI
2 = FLAT POSITION ONLY
4 = A-C AND D-C ROD

E-7018

NUMBERING

COATING COLOR

DOT

END SPOT

Color of spot or dot on stick electrode is for quick identification. End spot can be seen when electrodes are in container.

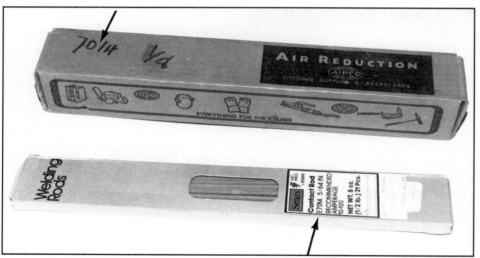

Although packaging and trade names may be different, look for American Welding Society (AWS) number (arrows). Shown are two packages of E-7014: Airco Air Reduction (top) and Sears Contact Rod.

To weld with E-6011 rod, hold a 1/8-in.-or-shorter arc. Move at a steady pace that is just fast enough to stay ahead of the molten slag. For welding overhead or vertical-up, reduce the current setting by one notch on the buzz-box.

E-6011 comes in 1/16-, 5/64-, 1/8-in. and larger sizes.

E-6013 is an excellent general-purpose welding rod for use with 220-volt a-c buzz-box welders when both easy operation and out-standing weld appearance are important. This rod can be used in all positions. The work must be cleaner than when welding with E-6011 rod. I recommend this 60,000-psi tensile-strength rod for projects such as building trailers.

To weld mild steel with E-6013 welding rod, drag the tip of the rod lightly against the work. Do not hold a gap as you would with E-6011 rod. Some companies call this a *contact rod* because you always keep it in contact with the base metal. Move steadily and just fast enough to stay ahead of the molten puddle. When welding sheet metal with E-6013, weld downhill. E-6013 welding rod comes in 1/16-, 5/64-, 1/8-in. and larger diameters.

E-7014 welding rod is also designed to be used with 220-volt a-c buzz-box welders. Compared to E-6011 and E-6013 rods, E-7014 is a slightly higher-strength welding rod with good appearance characteristics. The first two numbers in the identification indicate 70,000-psi tensile strength. This rod is commonly used for sheet-metal welding.

Another contact rod, you should lightly drag the tip of the rod on the base metal when laying a bead. Therefore, you don't have to maintain a gap with the arc. This is a good welding rod for beginners, but so is E-6011. E-7014 comes in 3/32-, 1/8-in. and larger diameters.

E-6010 welding rod, or *5P* as it is called by oil-field pipe welders, is an all-purpose, deep-penetrating welding rod for use with 220/440-volt d-c welders. It works similar to all-purpose a-c/d-c E-6011 rod, but leaves a much smoother weld and produces almost no spatter. It will work on dirty, oily or rusty pipe and other steel. E-6010 welding rod should not be used with an a-c welding machine. Set the machine for *d-c, positive or reverse polarity* when welding with this rod. It comes in 1/8-in.-diameter and larger sizes.

E-7018 LO-HI, sometimes called *low hydrogen,* should be used with 220/440-volt d-c arc-welders. This welding rod was developed originally for 70,000-psi tensile-strength, X-ray- quality welds in the nuclear-power industry. Regardless, the cost of E-7018 rod is similar to other welding rod, so it is one of the

Stick-electrode carrier keeps rods organized, clean and off floor. After finishing welding for the day, I store rod in sealed container to minimize moisture damage.

STICK-ELECTRODE WELDING ROD FOR CAST IRON

Coating Color	End Mark	Spot Color	AWS Rod Number	Use For Welding	Polarity + = Reverse − = Straight
Light Tan	Orange	—	E St	Cast-iron car parts, farm equipment	d-c + or a-c

Refer to page 85 for tips on how to weld cast iron.

STICK-ELECTRODE WELDING ROD FOR STAINLESS STEEL

Coating Color	End Mark	Spot Color	AWS Rod Number	Use For Welding	Polarity + = Reverse − = Straight
Gray	Yellow	—	E-308-16	Stainless steel 308 series	a-c or d-c d-c +
Gray	Yellow	Blue	E-347-16	Stainless steel 347 series	a-c or d-c d-c +

Refer to page 85 for welding stainless steel.

most commonly used welding rods. It produces high-quality, good-looking welds suitable for pipe welding that must be certified. E-7018 rod also works well for trailer frames, race-car frames and mild steel.

The work must be thoroughly cleaned and prepared. If you are sloppy, it is easy to bury slag pockets with E-7018 rod. If the metal is clean and properly prepared, the slag will actually peel off as the weld cools. This rod is highly susceptible to moisture damage, so it must be kept dry and clean at all times. Once you get the hang of using E-7018 rod, you'll like it.

EST and Lincoln Ferroweld are steel welding rods for making high-strength welds in cast iron when no machining is required afterward. These welding rods are used with 220-volt a-c welders or 220-volt d-c welders set to reverse—*positive*—polarity.

Maintain a short arc as you would with an E-6011 rod—don't touch the work with the rod tip. It's best to preheat the entire part to 400F (204C) prior to welding to minimize stresses. Cast iron should be welded in short, 1-

in.-long beads and allowed to cool in between. This reduces heat buildup and the likelihood of cracking.

E-308-16 stainless-steel rod may be used with a 220-volt welder for certain stainless-steel welds where weld appearance is not critical, such as pipe welding. It leaves a finish similar to E-6011

rod—a lot of spatter. Use a-c current or d-c current set to positive polarity.

Other specialty welding rod, designed for use in electric arc-welding machines, can be found at your local welding shop:
• Aluminum
• Bronze
• Hard surfacing—a very hard

STICK-ELECTRODE WELDING ROD FOR ALUMINUM

Coating Color	End Mark	Spot Color	AWS Rod Number	Use For Welding	Polarity + = Reverse − = Straight
White (lumpy)	—	—	AL-43	Aluminum plate and forgings	d-c + (only d-c)

Refer to page 85 for welding aluminum.

STORING ARC-WELDING ROD

It's essential that coated welding rod be kept dry and clean when stored. The two biggest enemies of *all* welding rod are moisture and dirt. Discard any coated welding rod that starts flaking. Following are two low-cost ways to store rod:
• An old, non-operating refrigerator with a 100-watt light bulb inside burning continuously is sometimes used to store large quanti-

ties of arc-welding rod. The heat of the light bulb dries the air in the refrigerator and the air-tight refrigerator keeps out humid ambient air.
• For small shops, sealable plastic containers are used. Four or five hours before the rod is to be used, it is removed from its container and oven baked at 200F (93C). This forces any moisture from the arc-welding-rod coating.

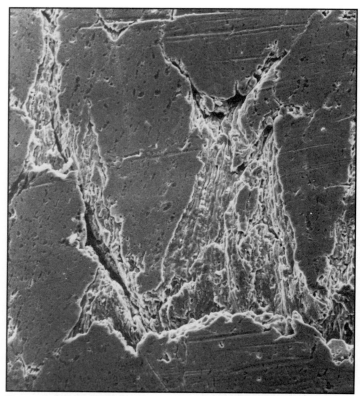

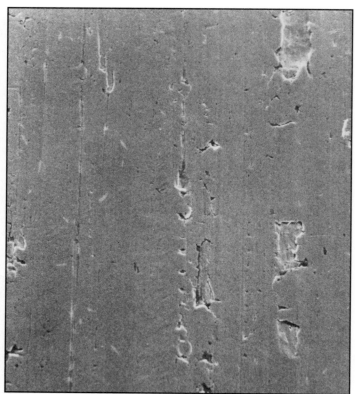

Commercial-grade TIG wire for 4130 steel magnified 1300 times is full of voids and surface imperfections. Compare to next micrograph that shows metallurgically-controlled wire; obviously a better product. Photo courtesy United States Welding Corporation.

1300X micrograph of high-quality welding rod shows surface tears from being drawn through a forming mandrel. Photo courtesy United States Welding Corporation.

BARE ROD FOR TIG WELDING

Number	Name	Use For	Method
Aluminum 4043	Aluminum rod	TIG-welding of most weldable aluminum, aircraft welding, race-car welding	TIG and Oxyhydrogen
Welco W1200	4130 bare rod	TIG-welding of mild steel and 4130 steel. Aircraft welding, race-car welding.	TIG and Oxyacetylene
308	Stainless steel rod	TIG-welding of 70% of all stainless steel. Aircraft welding, race-car welding.	TIG
SFA 5.16 ERTI-2	Titanium bare rod	TIG-welding of titanium with back gas and trailing shield or in vacuum chamber	TIG

Refer to Chapter 9 for how to TIG-weld metals listed above.

coating of steel applied over mild-steel surfaces where abrasive wear occurs, such as bulldozer blades, plow blades and steam-shovel scoops.

TIG WELDING

4043 is an uncoated aluminum welding rod for TIG-welding aluminum. I use this welding-rod alloy for most TIG-welding jobs on aluminum. It is available in several diameters: 0.020, 0.040, 1/16, 3/32, 1/8 and 5/32 in. You can weld the following aluminum alloys with 4043 welding rod: 1100, 5052, 6061, and 356 (casting).

Welco W-1200, AWS A5.2-69, Class RG60 is the most common TIG-welding rod for welding steel. For some reason, this rod is not listed in most welding-rod catalogs! I discovered it because I've always relied on my welding supplier to provide the best rod available. If the box of welding rod has a *heat number*—a reference number from the manufacturer's quality-control department—the particular batch of rod has been checked for quality. If a question comes up about the rod for one reason or another, the manufacturer can trace its history through this number. A heat number usually means the welding rod is for low-carbon, high-strength steel such as 4130.

Don't let anybody sell you copper-coated rod for TIG welding. The copper coating can cause blowholes in a weld and generates fumes that can be hazardous to your lungs.

If you absolutely insist, you can order 4130 welding rod from

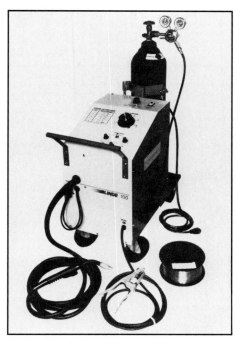

Linde 100 wire-feed welder with CO_2 cylinder and wire spool. Wire must be kept clean and dust-free. Photo courtesy Linde Welding Products.

WIRE ELECTRODE FOR MIG-WELDING STEEL	
Electrode	Application
E70S-1	Use when weld quality is not critical.
E70S-2	Good for welding dirty or rusty metal. Produces good weld quality in this application.
E70S-3	Use for single-pass welds where weld quality is not critical.
E70S-4	Similar to -3, but has higher silicon content.
E70S-5	Contains aluminum and high silicon content as deoxidizers for purifying weld. Use with high amperage on thick materials.
E70S-6	Use when high-impact resistance is needed.

S indicates solid-core wire. T in place of S indicates flux-core wire.

some dealers. But the chemists tell me that when you melt mild steel into low-carbon steel, as when welding 4130, the filler rod takes on the properties of the base metal. Consequently, the weld becomes as strong as the base metal. Mild steel has a tensile strength of 60,000 psi; 4130 tensile strength is a bit higher at 70,000 psi. So, it's easy to see that mild-steel rod should be OK for most 4130 TIG-welding applications. This mild-steel rod comes in 0.030-in., 1/16-in., 3/32-in., 1/8-in. and larger diameters.

308 stainless-steel rod is what's needed for TIG-welding most stainless-steel alloys. There are more than 80 different stainless-steel alloys; 90% of them are welded with 308 rod! Contact your dealer for special cases, or if you have any questions about which rod to use.

Stainless rod is non-magnetic, of course, and is identified with a small white tag taped to each rod with the identification number on it. This 308 stainless-steel rod is used for welding such things as race-car and airplane exhaust systems, kitchen cookware and missile parts.

SFA 5.16 ERTI-2 titanium rod is for TIG-welding one titanium alloy. The rod must *exactly* match the alloy. Another titanium alloy, Ti-6A1-4V—commonly referred to as *6-4*—is the one most commonly used because it's considered the "4130" of titanium. To weld 6-4 titanium, a 6-4 titanium rod must be used.

Titanium rods come in the same sizes as stainless rods. They are also identified with paper tags and usually must be special-ordered from most welding-supply shops.

Magnesium rod is not carried by most welding-supply shops. Because of its limited use, it's a special-order item. Use the same rod as the magnesium alloy being welded, such as AZ92A, AZ101Z or AZ61A rod. The best rod for TIG-welding *wrought* magnesium is AZ61A rod.

WIRE-FEED (MIG) WELDING

Electrodes used for MIG welding are smaller than those used in other types of welding. This is because of the high current and speed at which the filler metal is introduced into the weld puddle. For instance, wire sizes start as low as 0.020-in. diameter and increase in steps of about 0.005 in. Average size is 0.045 in. with a normal maximum of 0.090 in. A maximum diameter of 0.125 in. has been used in heavy industrial applications. Note: A smaller electrode will achieve more penetration at the same amperage.

E4043, E4146 and E5183 aluminum-wire alloys are available for MIG-welding aluminum. E4043 is most common for shop-welding projects, and for building aluminum trailers for heavy-duty, freight-hauling trucks.

E70S-1, -2, -3, -4, -5 and -6 steel-alloy wire is for welding mild steel. Prefix E indicates electrode. The second digit (7) refers to tensile strength in 10,000 psi, or 70,000 psi. The third digit (0) refers to position—horizontal in this instance. S means solid electrode, versus hollow core. A T instead of an S indicates hollow-core electrode. Dash numbers indicate the chemical composition of the wire as specified by the American Welding Society.

All numbers indicate varying percentages of carbon (C) and silicon (Si) except for -2. The E70S-2 wire also contains titanium (Ti), zirconium (Zr) and aluminum (Al). The higher the dash number, the higher the silicon content. See above chart for applications of each E70S filler.

13

Spot Welding

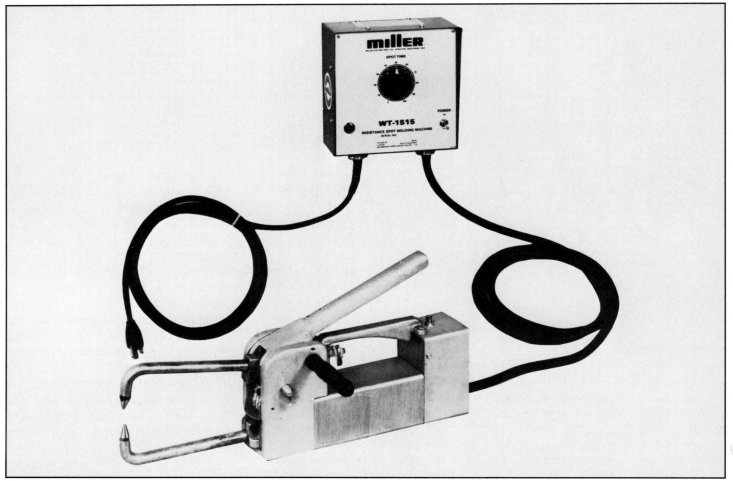

Hand-held spot welder with timer is ideal for sheet-metal shops. Several *tong*—jaw—lengths and tip types are available. Photo courtesy Miller Electric Mfg. Co.

Used strictly for joining one sheet-metal panel to another, spot welding is one of the oldest production-welding methods. The primary function of a spot welder is to make many welds with little effort and in a short time. It is also clean and requires no filler. Spot welding has been used extensively since the introduction of *unibody*—unitized body/frame—cars. It is also the standard welding procedure used by most sheet-metal fabricating industries.

This is one welding technique that can be understood and done without first learning how to gas weld. All that's necessary to produce a spot-weldable part is to have two clean pieces of sheet metal that will lay flat against each other at the weld joint. You cannot spot-weld a butt joint or T-joint because these joints lack sufficient surface area to clamp the two together. Therefore, all spot welds use lap joints.

In many cases, a lap joint is made by forming what is termed a *weld flange*. For example, if a panel butts into the side of another panel, as is the case of a T-joint, a flange about 1/2-in. wide is formed at the butting edge of the panel to provide an overlap—lap joint—with the other panel. This overlapping flange, or weld flange, can then be spot-welded.

You see such flanges every time you look under a car. It's the flange that runs lengthwise of the rocker panel, joining it to the floorpan and inner rocker panel.

TYPES OF SPOT WELDERS

Spot welding, more appropriately called *resistance welding,* uses pressure and electrical resistance through two metal pieces as the main ingredients of the weld. Heat is produced by high amperage routed through two mating workpieces clamped between two electrical contact points. No filler, flux or shielding is used. Because the pieces must be clamped together, access to both sides of

Spot welder in sheet-metal shop is portable type mounted on pedestal. This 220-volt, a-c Dayton unit is non-adjustable, with one setting for thin-gage sheet metal. Two sheets of steel are placed between jaws. When foot pedal is depressed, jaws clamp sheet metal between copper tips and apply electrical current. Electrical-resistance heating melts sheet metal, fusing them together at single spot.

Miller pedestal-mounted spot welder is water-cooled and air-operated, with ten-step heat control and clamping pressure adjustable up to 600 lb. Foot pedal operates spot welder. Photo courtesy Miller Electric Mfg. Co.

the joint normally is required. Heating is local, usually a 1/4-in.-diameter spot.

Spot welding can be done to almost all steel and aluminum alloys, but equipment cost varies widely. Whereas a spot welder capable of welding steel costs less than a buzz-box arc welder, a spot welder for welding aluminum usually costs more than $50,000!

Although spot welding is simple by itself, complete spot-welding setups can get extremely complicated. The degree of complexity depends on how the welding unit is mounted. For instance, robot-operated spot welders are in wide use, particularly in the auto industry. However, for the small fabrication or home shop, a pedestal-mounted or portable spot welder is the norm. The portable unit can be taken to the work, but the pedestal-mounted spot welder requires that you take the work to the machine.

Regardless of the mounting, most spot welders operate on single-phase a-c power. A *step-down* transformer converts the power-line voltage to about 250 volts.

Single-Phase Resistance—The standard, sheet-metal-shop spot welder is a single-phase resistance-type machine. With this type, a-c current is passed to two copper electrodes that clamp two thin pieces of metal together. The resulting *short circuit* causes the metal to heat to the melting point where the two electrodes meet.

This melts and forces the panels together locally, or at a *spot*.

Three-Phase Rectifier—This welder consists of a three-phase step-down transformer with diodes connected to the secondary circuit. These water-cooled, silicone diodes are connected in parallel. Current is also passed through two clamping electrodes.

Capacitor Discharge—With the capacitor-discharge welder, which also uses clamping electrodes, a bank of capacitors is charged from a three-phase rectifier and then discharged into an inductive transformer. It can be used to spot-weld dissimilar metals or delicate electronic parts.

Single Electrode—This is a simple unit that looks like a pistol and relies on a portable arc welder for its power source.

Unlike other spot-welder designs, the sheet-metal panels are not clamped together. This design feature makes it suitable for "skinning" sheet-metal panels on automobile bodies where access to the inner panel is blocked. The primary appeal of this light-duty spot welder is for the auto hobbyist.

Aluminum Spot Welder—If you try welding aluminum with a spot welder meant for welding steel, it will only burn the metal. And, it won't fuse the pieces together. Aluminum spot welders are sophisticated devices programmed to slowly apply pressure to the spot weld while gradually increasing voltage and current. They develop up to 76,000 amps to spot-weld 1/8-in.-thick aluminum.

Other Spot Welders—There are several other types of resistance welders, but most are high-production factory units, unsuitable for small workshops. For instance, projection welding utilizes small dimples, or projections, in the sheet metal. These projections arc as they touch another panel to complete the welding circuit, *flash-butt welding* the two together. A high electrical resistance at the projections is created, causing the panels to fuse together.

Another type of resistance weld is called *seam welding*. Two pieces of sheet metal are drawn between two rollers while pressure and electrical current are applied to the knurled rollers. This action provides for a *continuous* resistance weld, which is leak-free. A common application is in the manufacture of automotive gas tanks. The top and bottom halves are continuously welded at a flanged seam.

SPOT WELDERS ARE COMPLEX

Most spot-welder designs are complex even though using them is relatively easy. Spot-welder settings for current applied, on-and-off cycle timing, electrode pressure, and the shape and condition of the electrodes all affect weld quality. Once the proper setup is achieved, the operator can make numerous welds of equal quality.

ALUMINUM SPOT WELDING

When spot-welding low-carbon steel, a good weld can be made with a generalized setup using a wide range of current adjustments and an even wider range of *clamp, current-applied, current-off, pressure-held,* and *pressure-released* cycles or steps. However, aluminum is much harder to spot-weld. It requires a precise clamping cycle and slow application of an initially low current, building up to a spike of high current, then tapering off to low current again. The clamping force is initially high to prevent arcing of this high-resistance metal, then backed off to prevent thinning of the aluminum, and finally increased as the weld cools. These steps must be carefully done by an accurate timing mechanism. And, as you may

recall from the TIG and MIG chapters, aluminum is highly susceptible to surface oxidation that adversely affects weld quality. Consequently, aluminum must be absolutely clean to spot-weld.

Also, the spot-welder contacts must be dressed—filed—often to keep them clean and to maintain electrode shape and size. Tip size is important. For example, when 31,960-psi pressure is applied to a 1/4-in.-diameter surface, reduction of this contact-patch size to 3/16-in. diameter will reduce the applied pressure to 17,978 psi—a 44% reduction! This will ruin the weld.

Because of the critical factors and complex machinery involved, aluminum spot welders can cost more than an entire welding shop equipped with MIG, TIG, arc and gas welders!

Not to worry, though. There are other ways to join sheet aluminum. For instance, at Aerostar Airplanes I used an aluminum spot-welding machine to weld two-piece landing-gear doors. When the spot welder broke down, we substituted rivets for the spot welds—rivets and spot welds are similar in strength. Although spot welding is usually faster and much less prone to *working*, or moving, of the two sheets of metal joined, solid rivets will work nearly as well.

WELDABILITY OF METALS BY SPOT WELDING

Almost any metal that can be fusion-welded can be spot-welded. However, oil, grease, dirt or paint on either or both of the parts to be spot-welded can have a negative effect on weldability. For this reason, it is imperative that you clean the sheet metal thoroughly prior to attempting spot welding.

Galvanized steel is a natural for spot welding. Galvanized coating was once used extensively in heating and air-conditioning ducting. It is now used in automobile car bodies for rust prevention. Galva-

nized steel spot-welds easily with no special preparation to the metal. However, you should use a good breathing respirator because it generates poisonous zinc-based gases.

Stainless steel can be spot-welded, but it should be silver-brazed or TIG-welded instead.

Nickel-plated steel, such as that used in car-body trim, can be spot-welded. Unfortunately, heat will discolor the plating.

Refer to the weldability chart on page 128. It can be helpful for determining which metals can be spot-welded.

Metal Coatings—Special *weld-through sealers* are used in the auto industry for rust protection. They prevent water from entering blind areas in the body through weld seams. The sealer is applied to the facing side of a weld-seam half, then the panels are put in place and the seam is spot-welded right through the sealer. This makes a watertight spot-weld seam.

HOW TO USE A BASIC SPOT WELDER

Most simple spot welders have four controls:
- On-off power switch.
- Foot pedal or hand lever to cause the two electrodes to come together and make the spot weld.
- Electrode contact-pressure adjustment, usually air pressure.
- Amperage and timer adjustment. As metal thickness increases, so does arm-contact time and amperage.

Before making final spot welds, cut out some test scraps of the same material, about 2-in. square. Hold them together with locking pliers while spot-welding.

Start off with low pressure and amperage. Metals that have high electrical resistance, such as aluminum, require less overall current, but more exact control of the cycle. Steel and other metals with low resistance require greater amounts of current. Keep raising the pressure and amperage until

When using portable, non-adjustable spot welder, operator releases foot-pedal pressure as soon as sparks fly. This spot welder is manually timed by operator rather than by adjustable electric timer. Note long tongs for increased throat depth. To increase accessibility, upper tong is offset. Photo courtesy Brevard Sheet Metal, Inc.

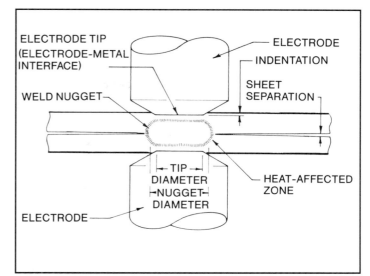

In spot welding, heat is produced by electrical resistance between copper electrodes. Pressure is simultaneously applied to electrode tips to force metal together to complete fusing process. Spot-weld-nugget size is directly related to tip size. Drawing courtesy The James F. Lincoln Arc Welding Foundation.

the spot is about 3/16—1/4-in. diameter.

For instance, a spot welder with 1/4-in. electrodes should be set to the following values to assure good welds in steel sheet: 9800 amps, 32,000-psi ram pressure. Clamp time and weld dwell time are programmed into the welder. Make a few spot welds using the suggestions just given and adjust the welder as required for good welds.

The distance between each weld is *pitch*. For most applications, weld pitch should be about 1 to 1-1/2 in. Judge the spacing as you move along a weld seam, making one weld after another. After making a few spot welds, perform the following tests.

SPOT-WELD TESTING

The ultimate method of checking spot-weld quality is destructive testing whereby the two spot-welded pieces are pulled apart. However, you can tell a lot about a spot weld by its appearance.

Good spot welds are essentially round. The depression matches the size of the spot-welder electrode-contact points, with only a slight amount of heat-affected metal around the spot.

MILD-STEEL SPOT-WELDING GUIDE		
Metal Thickness (in.)	Electrode Pressure (psi)	Current (amps)
0.010	200	4,000
0.020	300	6,000
0.030	400	8,000
0.040	500	10,000
0.050	650	12,000
0.062	800	14,000

Refer to chart when setting up a spot welder. Make adjustments as necessary.

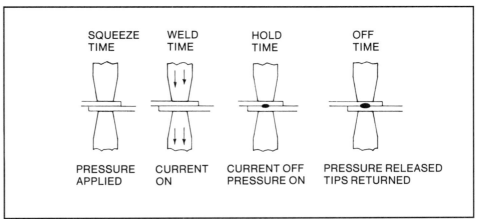

Four primary cycles of a spot weld are: squeeze, weld, hold and release. Pressure and current must be precisely controlled to assure good welds. Drawing courtesy The James F. Lincoln Arc Welding Foundation.

	Aluminum	Stainless Steel	Brass	Copper	Galv Steel	Steel	Monel	Tin	Zinc
METALS THAT CAN BE SPOT-WELDED									
Aluminum	X							X	X
Stainless		X	X	X	X	X	X	X	
Brass		X	X	X	X	X	X	X	X
Copper		X	X	X	X	X	X	X	X
Galv Steel		X	X	X	X	X	X	X	
Steel		X	X	X	X	X	X	X	
Monel		X	X	X	X	X	X	X	
Tin	X	X	X	X	X	X	X	X	
Zinc	X		X	X					X

X Indicates combinations that can be spot-welded.

The amount of indentation in the weld should be slight.

Bad spot welds look much different. Severe discoloration around a spot weld indicates overheating. Either the current was too high or the *dwell time*—duration—of the current was too long.

Little or no contact spot indicates a poorly fused spot weld. The current was too low or current dwell time too short. Spitting and sparking around the spot weld indicate one or more problems. Either the metal was dirty with paint, oil or grease, or the current was much too high for the metal thickness. It is also possible that clamping pressure was insufficient.

As with anything, it is easier to make corrections to spot-welding machine settings when you know what caused the problem. Be sure to analyze the first two or three spot welds before continuing to make welds.

Destructive Testing—The most common way to check the operation of a spot welder is to perform a pull test. Spot-weld two metal test strips together. Start by visually inspecting the spot weld; then tear the two strips apart. Look at the torn apart spot weld to determine whether it is good or bad. The amount of force it takes to

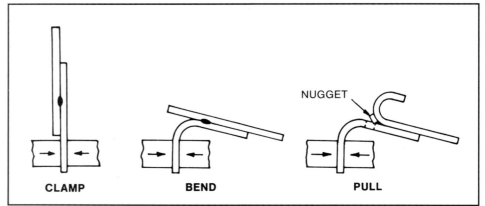

Test spot weld by clamping one piece in vise and *bending* two pieces apart. If spot weld pulls apart, penetration is poor. If spot weld pulls a *nugget*—spot-welded metal—from one piece, penetration should be good. **Drawing courtesy The James F. Lincoln Arc Welding Foundation.**

tear the strips apart is the best indicator. If you have to really wreck the metal to tear the weld apart, the weld is probably good.

To tear the strips apart, clamp one strip in a vise. With pliers or Vise Grips, peel the other strip away from the first strip. After you have torn the strips apart, look at the weld area. Usually, there will be a hole in one piece of metal and a weld *nugget*—fused metal spot—stuck to the other piece. *Pulling* a nugget indicates the weld was stronger than the base metal.

Check the size of the nugget. It should be almost the size of the face diameter of the electrode.

Check the shape of the nugget. It should have some of the metal from the other test strip attached to it. Look for brittle, crystalline-looking metal in the nugget. Brittleness is usually caused by excess heat, not enough pressure, or too much heat time. Look at the nearby drawings for how to do a spot-weld pull test.

USING A SINGLE-POINT SPOT WELDER

If you are restoring an antique car, or building a new car and don't want to use rivets, a hand-held spot welder such as the Braze'N'Spot Welder from The Eastwood Company, Malvern,

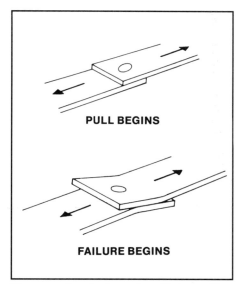

You can also test spot weld by *pulling* apart spot-welded strips in test rig. Drawing courtesy The James F. Lincoln Arc Welding Foundation.

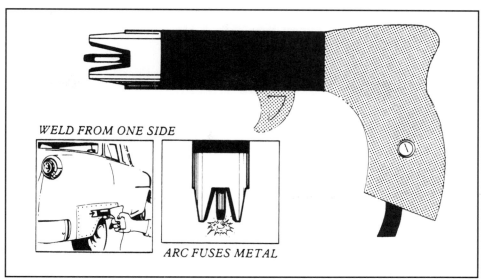

Hand-held spot welder uses 110-volt arc welder as power source to spot-weld 18—26-gage steel (0.050—0.018 in.). It can be used in place of rivets in body repairs. Drawing courtesy The Eastwood Company.

PA, may give you the desired results. The advantage of this type of spot welder is that you can weld panels where access to the back side is restricted. Eastwood's spot welder operates off a 50-amp buzz-box/transformer and plugs into 110-volt house current.

To use it, first ground the welder to the work. Next, push the welding gun against the outer panel, forcing it against the inner. Do this while you retract the carbon electrode by pulling the gun trigger. To make the spot weld, apply current and heat to the outer panel by releasing the electrode trigger. A spring will force the electrode against the outer panel, allowing amperage to heat the metal to a molten puddle. This should fuse the outer panel to the inner.

When the metal glows red to white hot, retract the electrode, but continue holding pressure on the gun and panel for 5 or 6 seconds until the hot spot cools. If everything was clean and fitted properly, you should have a good spot weld.

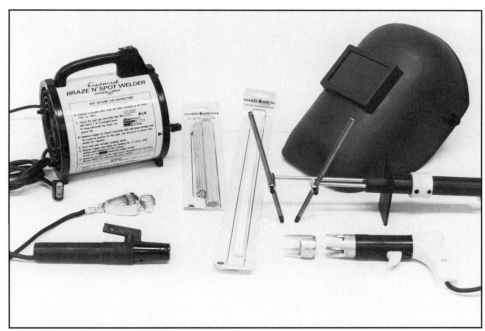

Complete Eastwood single-electrode spot-welder outfit includes 50-amp powerpack, arc- and spot-welding attachments, Chem-Arc brazing torch, welding rods, helmet and instructions. Photo by Ron Sessions.

14

Certification & Welding Professionally

Welding on projects such as this bridge may not require certification, but it won't hurt. On the other hand, some welding jobs do require certification, such as in nuclear power plants, aircraft industry and pipeline construction.

CERTIFICATION

Certification, or *qualification* as it's sometimes called, is very helpful if you plan to make a living as a weldor. However, you can weld professionally without being certified. Certification is something like having a college degree in sales. You can earn a living selling without it, but being certified may help you get a job by proving you are qualified. Regardless of certification, most large businesses and agencies conduct their own certification tests.

Documentation—What do you have to show once you're certified? You'll be given a wallet-size card listing the metals you are certified to weld and by what process(es). Also indicated will be whether you are a Class A (excellent), Class B (good) or Class C (OK) welder. Typically,

the card and certification expire in three, six or 12 months. You must be retested to stay certified.

You can be certified in different specialized areas: spot-welding contacts on home-heater thermostatic controls, TIG-welding stainless-steel pipe in the horizontal position in nuclear power plants, and so on. Two well-known certifications are for petroleum pipe-welding and aircraft TIG-welding. Certification testing is always conducted under strict supervision.

PIPE-WELDING CERTIFICATION

The exam for this certification process is welding a section of 5/16-in. wall, 6-in.-OD pipe called a *coupon*. The ends of two 6-in.-long sections of pipe are beveled in a lathe and placed end-

to-end for tack-welding. Usually, a *backing ring* is placed at the pipe joint with 1/8-in. spacers sticking out to obtain the proper *root opening*—distance between the pipe ends at the weld seam—to ensure that 100% penetration is obtained.

The pipe is placed at the weldor's eye level. This positions the weld seam so all welding positions—horizontal, vertical and overhead—and combinations of each must be used to complete one pass. The weldor starts with a *root pass*—the first bead. More than one pass is required to fill the weld seam. Except for the last one, following passes are appropriately called *fill passes*—usually one or two are required. The last pass is called the *cap,* or *weld-out pass.*

Slag must removed after each pass to eliminate possible slag inclusions.

After welding is completed, a coupon from the weld seam is removed and scrutinized. Coupons are cut from the pipe with a saw or cutting torch. The weld bead is then ground flat and even with the pipe surface. Each coupon is placed in a test machine and bent backward to a horseshoe shape. Any porosity or cracking of the weld or base metal will fail the weldor.

When certifying for arc welding, 5P welding rod is used for oil-field mild-steel pipe; E-7018 rod for 4130 steel pipe in steam and nuclear power-plant welding. TIG and wire-feed techniques are also used. As mentioned, the employer usually gives the certification test. However, there are also individual testing laboratories. One such laboratory is:

Advanced Testing Laboratories
4345 E. Imperial Hwy.
Lynwood, CA 90263

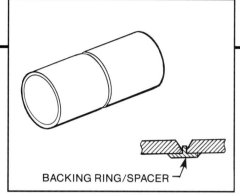

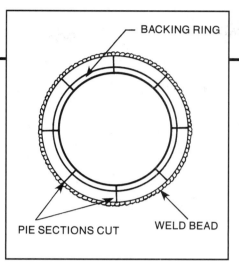

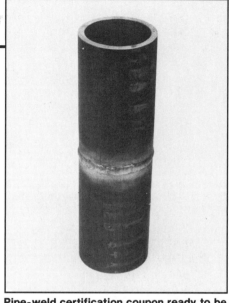

Pipe-welding certification starts with two pieces of beveled pipe mated end-to-end, sometimes with a backing ring and 1/8-in. spacers.

After pipe is welded, it is cut apart in sections—coupons. These coupons are tested for strength. Weld is sometimes X-rayed.

Pipe-weld certification coupon ready to be X-rayed and/or sectioned for destructive testing. Photo by Tom Monroe.

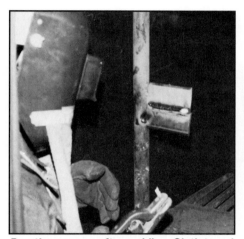

Practice coupon after welding. Cloth taped to weldor's helmet protects his neck against UV burns. Photo courtesy St. Phillips College.

After cleaning coupon of flux with chipping hammer, coupon is cooled in water tank.

Quick look at first weld bead indicates a rough, poorly penetrated weld. This student needs more *practice!*

AIRCRAFT-WELDING CERTIFICATION

Several coupons—both pipe and sheet—must be welded successfully before you can become a Class A certified aircraft weldor. Exact procedures may vary from country to country, depending on that government's regulatory laws. But basically, there are four groups of metals to be certified on:

- **Group I, 4130 Steel**—One cluster, cross-sectioned, polished and then etched for examination.
- **Group II, Stainless Steel**—Butt-weld 0.032 and 0.063-in. plates together. Complete fusion is required. Weld is sectioned, ground and bent. Visual inspection follows.
- **Group IV, Aluminum**—Butt-weld 0.032 and 0.063-in. plates together. Complete fusion is required. Weld is sectioned, ground and bent. Visual inspec-

tion follows.
- **Group VI, Titanium**—Special request; test production part by sectioning, polishing and 10X magnification inspection.

Group I, 4130-Steel Cluster Test Procedure—Test consists of TIG-welding three tubes to a flat plate with specified angles, dimensions and weld seams. The plate is 1/8-in. thick X 8-in. X 6-in. 4130 *condition N—normalized,* or stress-relieved. The three 1-in. OD X

Back side of weld exhibits excessive drop-through. Backing strip would help cure problem.

After instructor thinks student has a good weld, coupon is cut into strips and ground smooth for bend-testing.

Coupons are bent in guided bend-test machine to stretch metal and check for cracks, tears or porosity in weld.

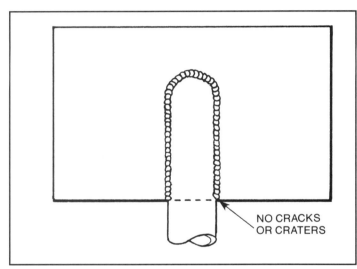

NO CRACKS OR CRATERS

After first tube is welded 100% on both sides, inspect for craters and cracks where tube and plate meet.

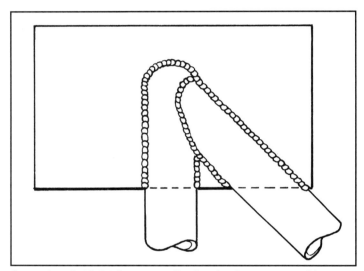

Second and third tubes must fit closely. Any gaps could cause failure.

0.063-in.-wall tubes are 6-in. long.

The sequence is to TIG-weld the first tube 100% all around on one side of the plate. After each weld pass, the torch should be held over the weld until argon-gas flow stops—usually 5—6 seconds. It is then allowed to cool in draft-free, still air for 20 minutes. The weld seam must fit tight because fitting is also graded.

After the first weld cools, scale is carefully removed from the other side of the plate and tube. Sandblasting should not be done due to the danger of embedding sand in the weld, causing the weld to fail the test. Inclusions similar to that caused by slag would result. Instead, scale should be removed with a stainless-steel wire brush. Be sure to clean the weld seam with MEK or alcohol prior to fitting and welding. Otherwise, the weld will be contaminated. Now, the back side of the first tube can be welded 100%.

Again, all scale is removed from the coupon and the second tube is fitted. The second tube should be filed so it fits tightly around the weld bead on the first tube. If too much filing is done, causing a large gap, the weld could fail the test. After the second tube is welded, the tubes and plate are cleaned as before. Weld the third tube like the second.

After completing the final weld, the coupon is cleaned and submitted to the lab for inspection. The testing laboratory will not grant certification if cracks, porosity, cratering, incomplete fusion or a poor fit is found.

Aluminum Butt Plates Test Procedure—Minutes prior to actually making the TIG weld, a Scotchbrite abrasive pad should be used to slightly roughen each piece of aluminum about 1/2 in. on both sides of the weld seam. The two pieces should be clamped together. Also, it's important that

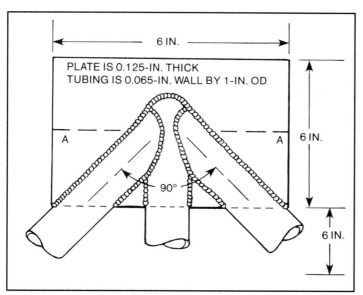

PLATE IS 0.125-IN. THICK
TUBING IS 0.065-IN. WALL BY 1-IN. OD

6 IN.
6 IN.
6 IN.
90°

Penetration on aircraft-weldor's certification test cluster must be at least 15% into 1/8-in. plate. Weld is not the only thing that's graded. Fitting tubes to specifications is also graded.

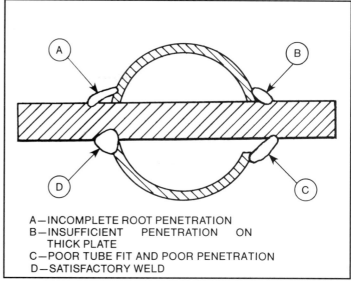

A—INCOMPLETE ROOT PENETRATION
B—INSUFFICIENT PENETRATION ON THICK PLATE
C—POOR TUBE FIT AND POOR PENETRATION
D—SATISFACTORY WELD

Common defects to avoid when fitting and welding aircraft-certification cluster.

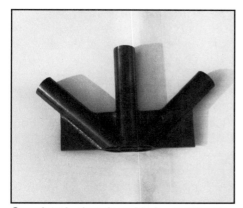

Completed aircraft-weldor's certification cluster makes great wall sculpture or desk ornament.

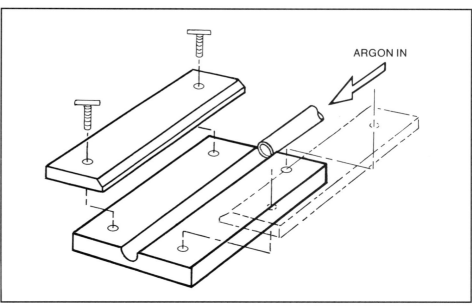

ARGON IN

To pass certification on stainless steel, back-gas and clamping fixture must be used.

a 1/4-in. air space be left under the weld seam. This ensures that 100% penetration is achieved. If the bead is not thicker than the base metal on *both* sides, you fail the test. Even though I don't use back gas, I use my stainless-steel back-gas clamping fixture to clamp the aluminum.

When the testing lab checks the aluminum weld coupon, they sand the weld bead to the thickness of the base metal, cut the coupon into strips and pull-test them. The aluminum strips must not break in the heat-affected area. Also, craters and porosity are not allowed.

Stainless-Steel Butt Plates Test

Procedure—Stainless-steel coupons are always TIG-welded with back-gas purging to prevent *crystallization* on the back side of the bead. Clamping and back-gas purging are done with the fixture pictured above.

Prior to welding, clean the stainless-steel plates with MEK or alcohol. Use clean, high-quality welding rod. To ensure that argon gas covers the weld while solidifying, keep the torch over the bead after completion until the purge-gas stops. The timer on the

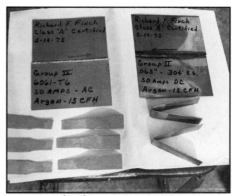

Pull-test (left) and bend-test coupons (right). Coupons before testing (top) and after testing (bottom). All passed class "A" specs.

133

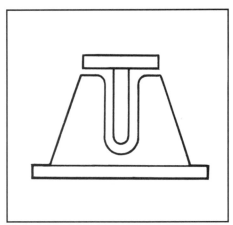

Fixture bends each coupon section.

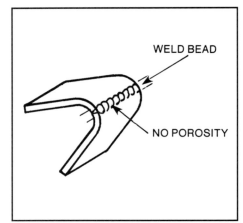

Cracks, tears or porosity in weld are not allowed. Failures due to flux or slag inclusions are most common defects.

TIG machine usually provides 5—6 seconds of gas flow after the torch is off.

The testing laboratory sands the stainless-steel weld down to base-metal thickness, then cuts it into test strips. The strips are then bend-tested at the weld seam. Without 100% penetration, the bend test will break through at the bead. If so, the weld will fail.

Welding Titanium—This is an on-the-job test that uses an actual production part. The part is then destruction-tested. It is sectioned through, pulled and bent, and subsequently polished and inspected with a 10X magnifying glass.

WELDING FOR A LIVING

I once saw a bumper sticker that read, "Welding holds the world together." Welding actually does hold more things together than most people realize. For instance, your automobile has hundreds of spot welds, seam welds and weld beads. Take a drive and you'll see steel bridges with thousands of welds. You'll pass by skyscrapers that are welded-steel structures underneath all that brick, mortar and glass. People like you and I welded those things together. Even robot welders have to be set up and adjusted by a weldor—er, uh, *welding technician*—so the welds are as good as those welded by human hands.

Ordinarily, welding pays as well as most skilled trades. In many cases, a weldor's salary matches that of a degreed engineer. Likewise, a welding engineer is one of the highest-paid engineers. The highest-paying jobs for welding engineers are in offshore structures and structural-steel industries. Why? The forces of supply and demand govern here. Welding is a skilled trade and there *is* an element of danger involved.

Welding is like riding a bicycle; once you learn how, you never forget. But you do need to practice from time to time. Once your friends and neighbors discover that you can weld, you'll be asked to fix many things—now that will keep you in practice. You could even make a little extra money.

WELDING SCHOOLS

Training—Some colleges and industrial-arts schools provide evening and weekend classes for weldors desiring to become certified. Trade schools also provide classes for welding certification. In these classes, a student is first taught the basics on pieces of scrap metal, and then given several weeks of nothing but practice, practice, practice to improve skills.

If you are serious about wanting to become a professional weldor, start by taking a welding class. Most high schools teach welding as an elective. And most two-year colleges provide a degree in welding technology. It's interesting to note, though, that many four-year colleges and universities do *not* provide welding courses or degrees. Those that do usually are engineering schools.

Most welding manufacturers, such as Lincoln Electric Co. and Hobart Brothers, operate welding schools at their factories. Lincoln's school, the James P. Lincoln Arc Welding Foundation, provides scholarship awards for students of other colleges and universities. For more information contact:

Aluminum Company of
America (ALCOA)
Technology Marketing Division
303-C Alcoa Bldg.
Pittsburg, PA 15219

Hobart Brothers
Box HW-34
Troy, OH 45373

The James P. Lincoln Arc
Welding Foundation
Box 17035
Cleveland, OH 44147

Projects

Welding projects can be as complex as aluminum-bodied stock car, designed by Tom Monroe, or as simple as welding cart, following page. Don't start complex project before you've developed necessary skills. Body panels are from Five Star Fabricating of Silver Lake, Wisconsin. Photo by Tom Monroe.

This chapter should be the most rewarding of all. You should've mastered one or more of the welding techniques covered in the book and you're now ready to put your equipment and new-found skills to work.

Each project is ranked by usefulness and degree of difficulty. The gas-welding cart is first because it doesn't have to be perfect, yet it will always be useful. The airplane is listed last because it *must* be perfect!

Projects are prioritized by the most-preferred to least-preferred welding processes. For instance, I explain that the gas-welding cart can be welded with oxyacetylene. But you can also braze it, arc-weld it, or even TIG- or MIG-weld it. Each project lists the preferred welding method for a beginner as first choice. Second and third choices are listed in the order I prefer, regardless of skill level.

PROJECT #1— GAS-WELDING CART

Welding Process—Oxyacetylene welding, brazing and cutting with a cutting torch. Arc, TIG and wire-feed welding are also acceptable.

Materials—Purchase the materials at a business that sells *steel*. Look in the phone-book yellow pages. Buy the wheels from a hardware or building-supply store.

Cut the Parts to Fit—With a cutting torch, cut the sheet metal to fit; cut the tubing and rod with a hacksaw. Heat the two long side tubes with a welding tip and bend to a 30° angle to form handles.

Tack-Weld—Start with the flat base and tack-weld the handles to the base. Keep everything square. Now, tack-weld all other parts into place.

Weld—You can gas-weld most of the cart. But, if you can't get a good puddle because of too-little flame and too-thick metal, braze or arc-weld it. Add hold-down

PROJECTS FOR WELDORS

Order of Skill Required	Name of Project	Preferred Welding Method
1.	Gas-Welding Cart	Oxyacetylene, or Arc, Brazing, Wire-Feed, or TIG.
2.	Push Cart	Oxyacetylene, or Arc, Brazing, Wire-Feed, or TIG.
3.	Mechanical Finger	Oxyacetylene, or Arc, Brazing, Wire-Feed, or TIG.
4.	Sander and Grinder Stand	Arc or Wire-Feed.
5.	Welding and Cutting Table	Arc.
6.	Jack Stands and Work Stands	Arc or Wire-Feed.
7.	Engine Stand	Arc or Wire-Feed.
8.	Go Kart	Oxyacetylene, TIG or Wire-Feed.
9.	Go Kart or Motorcycle Trailer	Arc or Wire-Feed.
10.	Hydraulic Press for Floor or Bench	Arc.
11.	Engine Hoist	Arc.
12.	Tow Bar for Cars, Jeeps	Arc and Wire-Feed or TIG.
13.	Trailers for Race Cars, Airplanes	Arc.
14.	Trailers, Utility	Arc.
15.	Experimental Airplane	Oxyacetylene or TIG.

Your first welding project should be a simple one, such as this welding cart. When finished, you'll also have something to carry and store your gas welder and accessories. Project should not take much more than two days.

Completed gas-welding cart has been put to good use.

straps or chain as shown in the photographs. Add some accessory hooks as desired. Paint to suit.

PROJECT #2—PUSH CART
Welding Process—Oxyacetylene welding or brazing. Cut with a cutting torch. Arc, TIG and wire-feed welding are also acceptable.

Materials—Buy the materials at a business that sells *steel*. Look in the phone-book yellow pages. Buy the wheels from a hardware or building-supply store.

Cut Parts to Fit—With a cutting torch, cut the sheet to proper size. Cut the tubing, strap steel, and axle shaft with a hacksaw. Heat the two side bars to bend them 30° and bend the handle with heat.

Tack-Weld—Start with the flat base and tack-weld the handles to the base. Keep everything square. Tack-weld the axle so wheels barely touch the floor with the cart standing up at rest.

Weld—Gas-weld or braze assembly. Do not try to weld a joint after brazing—it won't work. Paint to suit.

PROJECT #3—
MECHANICAL FINGER
Welding Process—Gas welding, brazing, TIG or wire-feed welding. Arc welding is OK, too.

Materials—Look in the phone-book yellow pages under **steel**. Get 2-ft lengths of 1/4, 3/8 and 1/2-in.-round rod.

PROJECT #4—
SANDER & GRINDER STAND
Welding Process—Weld with an arc welder and cut with a cutting torch.

Materials—Buy the materials at a business that sells *steel*. You may also be able to pick up some of the materials at a scrap-metal yard.

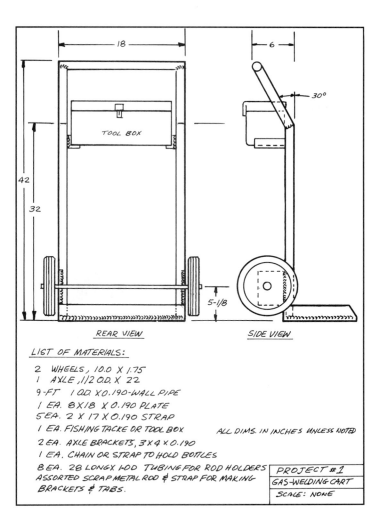

REAR VIEW   SIDE VIEW

LIST OF MATERIALS:

2 WHEELS, 10.0 X 1.75
1 AXLE, 1/2 O.D. X 22
9-FT 1 O.D. X 0.190-WALL PIPE
1 EA. 8 X 18 X 0.190 PLATE
5 EA. 2 X 17 X 0.190 STRAP
1 EA. FISHING TACKLE OR TOOL BOX   ALL DIMS. IN INCHES UNLESS NOTED
2 EA. AXLE BRACKETS, 3 X 4 X 0.190
1 EA. CHAIN OR STRAP TO HOLD BOTTLES
8 EA. 28 LONG X 1.OD TUBING FOR ROD HOLDERS
ASSORTED SCRAP METAL ROD & STRAP FOR MAKING
BRACKETS & TABS.

| PROJECT #1 |
| GAS-WELDING CART |
| SCALE: NONE |

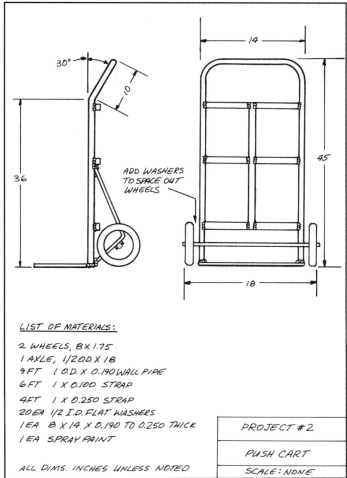

ADD WASHERS
TO SPACE OUT
WHEELS

LIST OF MATERIALS:

2 WHEELS, 8 X 1.75
1 AXLE, 1/2 O.D. X 18
9 FT 1 O.D. X 0.190 WALL PIPE
6 FT 1 X 0.100 STRAP
4 FT 1 X 0.250 STRAP
20 EA. 1/2 I.D. FLAT WASHERS
1 EA. 8 X 14 X 0.190 TO 0.250 THICK
1 EA. SPRAY PAINT

ALL DIMS. INCHES UNLESS NOTED

| PROJECT #2 |
| PUSH CART |
| SCALE: NONE |

Second project can be gas- or arc-welded. Handy push cart can be used to move washing machines, refrigerators, garbage cans or whatever.

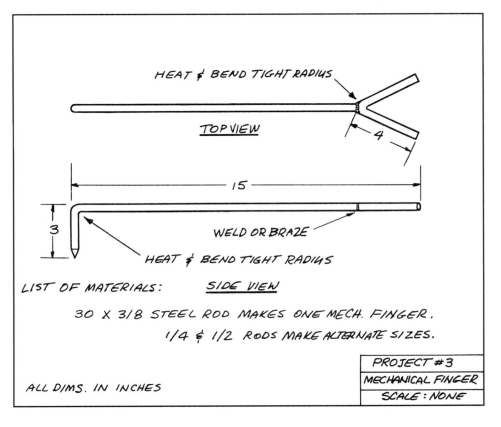

HEAT & BEND TIGHT RADIUS

TOP VIEW

4

15

3

WELD OR BRAZE

HEAT & BEND TIGHT RADIUS

LIST OF MATERIALS: SIDE VIEW

30 X 3/8 STEEL ROD MAKES ONE MECH. FINGER.
1/4 & 1/2 RODS MAKE ALTERNATE SIZES.

ALL DIMS. IN INCHES

| PROJECT #3 |
| MECHANICAL FINGER |
| SCALE: NONE |

137

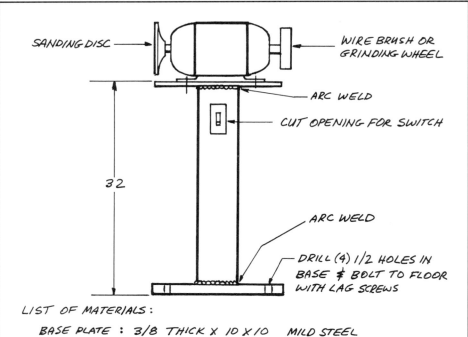

SANDING DISC

WIRE BRUSH OR GRINDING WHEEL

ARC WELD

CUT OPENING FOR SWITCH

32

ARC WELD

DRILL (4) 1/2 HOLES IN BASE & BOLT TO FLOOR WITH LAG SCREWS

LIST OF MATERIALS:

BASE PLATE : 3/8 THICK X 10 X 10   MILD STEEL
TOP PLATE : 1/4 THICK X 8 X 12   MILD STEEL
LEG : 4-DIAMETER PIPE (ROUND OR SQUARE)
SWITCH : SINGLE-POLE LIGHT SWITCH BOX & COVER
MOTOR : COMMERCIAL GRINDER MOTOR OR MAY BE MADE FROM 1725-RPM CLOTHES-DRYER MOTOR.

| PROJECT # 4 |
| GRINDER STAND |
| SCALE : NONE |

ALL DIMS. IN INCHES

You can build this grinder stand with 3-ft section of round or square tubing and some flat plate. Sander/grinder or wire-brush stand takes about four hours to build. Electric motor is from a clothes dryer.

Stands similar to that for grinder support anvil and vise. Both use heavier material and larger bases for stability. Note I-beam used for vise stand.

Cut Parts to Fit—With a cutting torch, cut the pipe and plates to size.

Weld—Arc-weld the plates to the ends of the pipe. Paint to suit.

PROJECT #5— WELDING & CUTTING TABLE

Welding Process—Cutting torch and arc welding.

Materials—Look in the phonebook yellow pages under **steel**.

Cut the Parts to Fit—This is a more advanced project, so do your own trimming to make the lower frame and secondary parts. Cut the legs to length. Make the top frame of 1-1/4 X 1-1/4-in. angle. *Make it square*. Weld on the 3/8-in.-thick table top. Trim all other parts to fit. Leave the sides open or fill them in with 16-gage (0.060-in.) sheet metal. You can make doors and a catch tray for sparks from the cutting torch.

Tack-weld the 1 X 1/4-in. straps in place so you can replace them

138

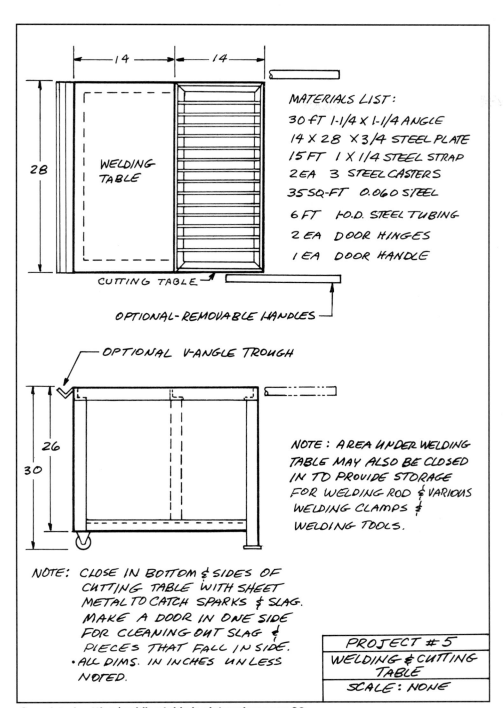

MATERIALS LIST:

30 FT 1-1/4 X 1-1/4 ANGLE
14 X 28 X 3/4 STEEL PLATE
15 FT 1 X 1/4 STEEL STRAP
2 EA 3 STEEL CASTERS
35 SQ-FT 0.060 STEEL
6 FT 1 O.D. STEEL TUBING
2 EA DOOR HINGES
1 EA DOOR HANDLE

NOTE: AREA UNDER WELDING TABLE MAY ALSO BE CLOSED IN TO PROVIDE STORAGE FOR WELDING ROD & VARIOUS WELDING CLAMPS & WELDING TOOLS.

NOTE: CLOSE IN BOTTOM & SIDES OF CUTTING TABLE WITH SHEET METAL TO CATCH SPARKS & SLAG. MAKE A DOOR IN ONE SIDE FOR CLEANING OUT SLAG & PIECES THAT FALL INSIDE.
• ALL DIMS. IN INCHES UNLESS NOTED.

PROJECT #5
WELDING & CUTTING TABLE
SCALE: NONE

Completed cutting/welding table is pictured on page 62.

Two stands at left are homemade; two at right are store-bought. Don't use tall stands to support car. They would be very unstable.

Work stand with pipe roller is handy for supporting large assemblies at work height in a welding shop.

easily when they get "ratty" from cutting-torch flame.

Paint everything but the table top.

PROJECT #6— JACK STANDS & WORK STANDS

Welding Process—Arc welding, cutting torch.

Materials—1-1/2—2-in. pipe for center shaft. Metal plate is 0.190—0.375-in. thick, depending on which of the pictured stands you make.

Plan Before Cutting—When you decide which work stand to build, make a simple drawing like the ones in this chapter. Don't just start cutting and welding. Other people may want you to make work stands for them once they see yours.

Sandblast & Paint—This prevents corrosion and improves the project's appearance.

PROJECT #7— FOLDING ENGINE STAND

Welding Process—Cutting torch and arc welding. TIG or wire-feed welding is OK.

Materials—Buy only the metal

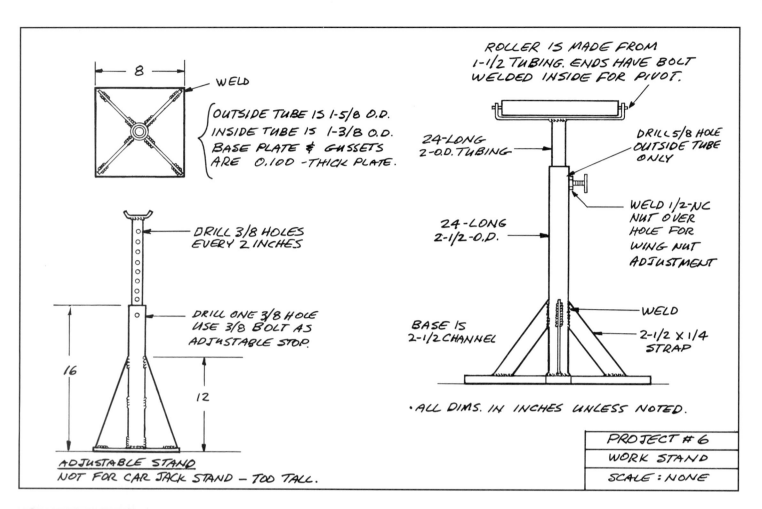

WELD

8

OUTSIDE TUBE IS 1-5/8 O.D.
INSIDE TUBE IS 1-3/8 O.D.
BASE PLATE & GUSSETS
ARE 0.100 -THICK PLATE.

DRILL 3/8 HOLES
EVERY 2 INCHES

DRILL ONE 3/8 HOLE
USE 3/8 BOLT AS
ADJUSTABLE STOP.

16

12

ADJUSTABLE STAND
NOT FOR CAR JACK STAND — TOO TALL.

ROLLER IS MADE FROM
1-1/2 TUBING. ENDS HAVE BOLT
WELDED INSIDE FOR PIVOT.

24-LONG
2-O.D. TUBING

DRILL 5/8 HOLE
OUTSIDE TUBE
ONLY

24-LONG
2-1/2-O.D.

WELD 1/2-NC
NUT OVER
HOLE FOR
WING NUT
ADJUSTMENT

BASE IS
2-1/2 CHANNEL

WELD

2-1/2 X 1/4
STRAP

· ALL DIMS. IN INCHES UNLESS NOTED.

| PROJECT # 6 |
| WORK STAND |
| SCALE : NONE |

Engine stand folds so it can be stored out of way or carried in car trunk.

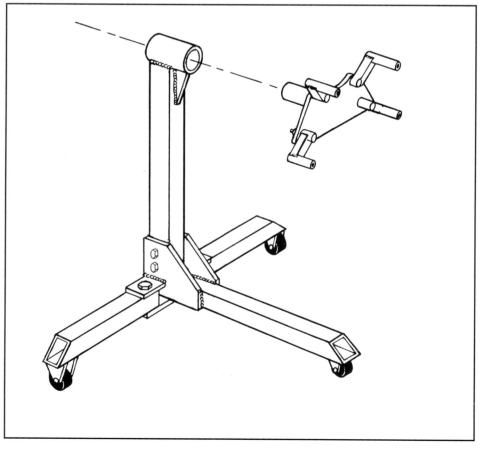

specified in this drawing, unless you want to change the design. The folding mechanism is semi-precision and won't work unless specified materials and dimensions are used. Buy from a steel-products warehouse. Material might have to be special-ordered. Go to your local hardware store for casters.

Cut Parts to Exact Size—Cut parts slightly *oversize* so they will be the correct size after grinding away the cutting-torch marks. Always cut an oversize radius on each square tube, then grind and file to shape.

Drilling Holes—Drill the 1/2-in. holes in the flat plates and square tubes *before* welding. Use the holes to jig the parts in place when welding.

Weld—Make sure everything is square! *First*, weld two 1/4-in. plates to the 28-in.-long tube. Temporarily install a vertical square tube when welding the plates. It will act as a welding fixture. Place a thin washer on one side of each tube that's to be bolted in place to ensure there'll be clearance after welding. Weld each seam 100%, *except* where the vertical post fits. With the post in place, you can see where not to weld. *Second*, weld the 7-in.-long plate to the bottom, and the two 2 X 2-1/4-in. plates to the sides of the 1/4-in.-thick gussets. *Third*, weld the 2-in. tube to the top of the 32-1/2-in.-long tube at a 3° angle so the engine tilts slightly nose up when on the stand. This makes the engine sit about right when it's on the stand—up to a maximum of about 700 lb, sufficient to support a small-block V8.

Finish welding the lock bolts in place and make the engine adapter plate. Install the casters. The nose caster should swivel and lock; side casters should be fixed.

Sandblast & Paint—The four bolts that thread into the engine block at the bellhousing flange should be minimum Grade-5; preferably Grade-8. The stand folds to a size of 7 X 8 X 40 in.

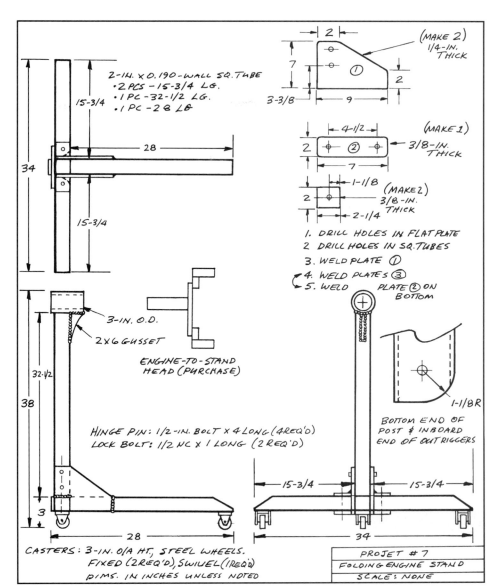

Although complex, go-kart project is relatively simple when broken into separate components, page 142.

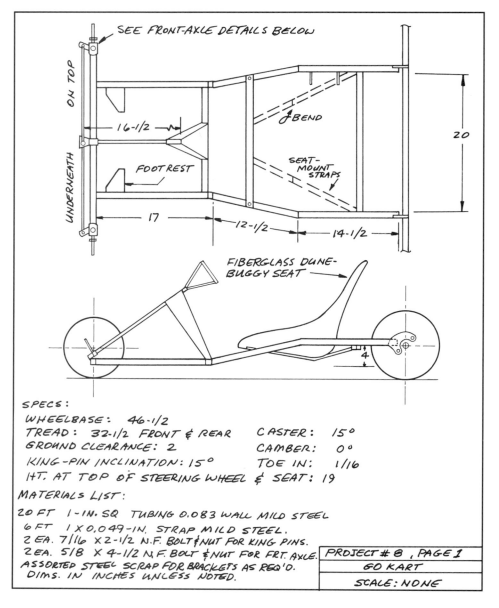

SEE FRONT-AXLE DETAILS BELOW

ON TOP

UNDERNEATH

16-1/2

BEND

20

FOOTREST

SEAT-
MOUNT
STRAPS

17

12-1/2

14-1/2

FIBERGLASS DUNE-
BUGGY SEAT

4

SPECS:
WHEELBASE: 46-1/2
TREAD: 32-1/2 FRONT & REAR   CASTER: 15°
GROUND CLEARANCE: 2   CAMBER: 0°
KING-PIN INCLINATION: 15°   TOE IN: 1/16
HT. AT TOP OF STEERING WHEEL & SEAT: 19
MATERIALS LIST:

20 FT 1-IN. SQ TUBING 0.083 WALL MILD STEEL
6 FT 1 X 0.049-IN. STRAP MILD STEEL.
2 EA. 7/16 X 2-1/2 N.F. BOLT & NUT FOR KING PINS.
2 EA. 5/8 X 4-1/2 N.F. BOLT & NUT FOR FRT. AXLE.
ASSORTED STEEL SCRAP FOR BRACKETS AS REQ'D.
DIMS. IN INCHES UNLESS NOTED.

| PROJECT # 8, PAGE 1 |
| GO KART |
| SCALE: NONE |

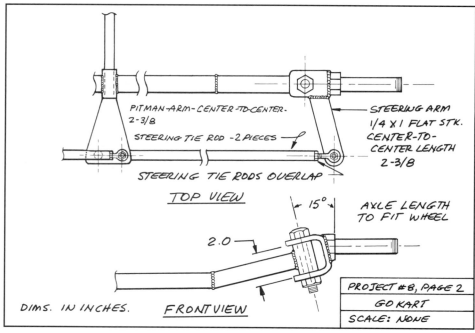

PITMAN-ARM-CENTER-TO-CENTER.
2-3/8

STEERING TIE ROD -2 PIECES

STEERING TIE RODS OVERLAP

TOP VIEW

STEERING ARM
1/4 X 1 FLAT STK.
CENTER-TO-
CENTER LENGTH
2-3/8

15°

AXLE LENGTH
TO FIT WHEEL

2.0

DIMS. IN INCHES.   FRONT VIEW

| PROJECT #8, PAGE 2 |
| GO KART |
| SCALE: NONE |

PROJECT #8— GO-KART

Welding Process—Oxyacetylene, but TIG welding gives best appearance and wire-feed is OK.

Materials—You'll probably have to order square tubing from a steel-supply house, but plate and scrap steel should be available locally. The tires, wheels, rear axle, sprockets, chain and engine will have to be ordered from a go-kart company. A 250cc engine is the maximum recommended size; 50cc is the smallest. Buy a fiberglass seat made for a dune buggy.

Cut Parts to Exact Size—Make a welding jig from 1/2-in. particle board or plywood. You'll need a 3 X 4-ft base. Draw a full-size frame plan, top view on the plywood or particle board. It should look just like the plans. Block up the rear frame 4-in. higher than the front. Fit the diagonal tubes last. You can use certain parts to line up other parts. A good example is using the steering shaft to line up the steering bracket.

Tack-Weld—Without burning up the jig board, tack-weld the complete assembly, then remove the frame from the board. Check the frame for warpage; twist as necessary to straighten it. Weld opposite joints in sequence as you go to minimize warpage.

Assemble—Because of variables in wheels, tires, axles, engines and other parts, you'll detail these attachments yourself. Don't paint anything until you've preassembled everything and have the kart ready to run, except for fuel in the tank. Correct any problems and take it apart. Then, paint it.

PROJECT #9— GO-KART OR MOTORCYCLE TRAILER

Welding Process—Arc welding, cutting torch also required. Wirefeed OK if heavy-duty machine is used.

Materials—Select a wheel-and-axle combination from a lightweight boat trailer, new or used. Buy the steel from a steel-products

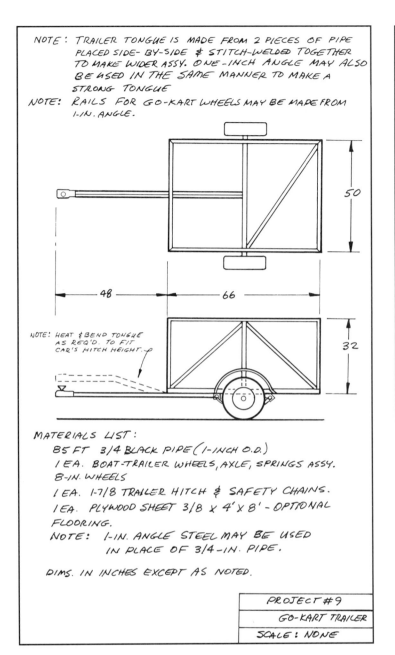

NOTE: TRAILER TONGUE IS MADE FROM 2 PIECES OF PIPE
PLACED SIDE-BY-SIDE & STITCH-WELDED TOGETHER
TO MAKE WIDER ASSY. ONE-INCH ANGLE MAY ALSO
BE USED IN THE SAME MANNER TO MAKE A
STRONG TONGUE

NOTE: RAILS FOR GO-KART WHEELS MAY BE MADE FROM
1-IN. ANGLE.

50

48   66

32

NOTE: HEAT & BEND TONGUE
AS REQ'D. TO FIT
CAR'S HITCH HEIGHT.

MATERIALS LIST:
85 FT 3/4 BLACK PIPE (1-INCH O.D.)
1 EA. BOAT-TRAILER WHEELS, AXLE, SPRINGS ASSY.
8-IN. WHEELS
1 EA. 1-7/8 TRAILER HITCH & SAFETY CHAINS.
1 EA. PLYWOOD SHEET 3/8 x 4' x 8' - OPTIONAL
FLOORING.
NOTE: 1-IN. ANGLE STEEL MAY BE USED
IN PLACE OF 3/4-IN. PIPE.

DIMS. IN INCHES EXCEPT AS NOTED.

| PROJECT #9 |
| GO-KART TRAILER |
| SCALE: NONE |

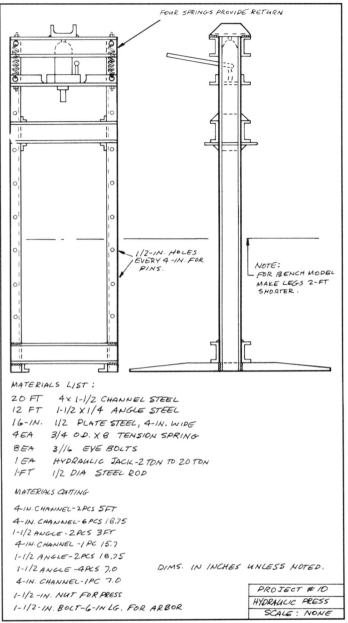

FOUR SPRINGS PROVIDE RETURN

1/2-IN. HOLES
EVERY 4-IN. FOR
PINS.

NOTE:
FOR BENCH MODEL
MAKE LEGS 2-FT
SHORTER.

MATERIALS LIST:
20 FT   4 x 1-1/2 CHANNEL STEEL
12 FT   1-1/2 x 1/4 ANGLE STEEL
16-IN. 1/2 PLATE STEEL, 4-IN. WIDE
4 EA.   3/4 O.D. X 8 TENSION SPRING
8 EA.   3/16 EYE BOLTS
1 EA.   HYDRAULIC JACK-2 TON TO 20 TON
1-FT   1/2 DIA STEEL ROD

MATERIALS CUTTING

4-IN. CHANNEL-2 PCS 5 FT
4-IN. CHANNEL-6 PCS 18.75
1-1/2 ANGLE-2 PCS 3 FT
4-IN. CHANNEL-1 PC 15.7
1-1/2 ANGLE-2 PCS 18.75
1-1/2 ANGLE-4 PCS 7.0
4-IN. CHANNEL-1 PC 7.0
1-1/2-IN. NUT FOR PRESS
1-1/2-IN. BOLT-6-IN LG. FOR ARBOR

DIMS. IN INCHES UNLESS NOTED.

| PROJECT #10 |
| HYDRAULIC PRESS |
| SCALE: NONE |

warehouse. The trailer hitch can be purchased from Sears, Wards or similar store.

Dimensions—Plans apply to a double-deck go-kart trailer. For motorcycles, adjust dimensions to fit specific motorcycles hauled.

Construction Method—Lay out and build the flat frame or floor first. Install the trailer tongue and hitch next, keeping the tongue square to the frame! Next, weld the spring perches and axle. Be sure the axle is square to the trailer frame. Next, add either the upper level for a go kart or ramps for motorcycles.

PROJECT #10— HYDRAULIC PRESS, FLOOR OR BENCH

Welding Process—Cutting torch and arc welder. Drill press or 1/2-in. portable drill also required.

Materials—Press can be made as a 5-ft floor model or a 3-ft bench model. For the bench model, shorten the legs 2 ft. The plans call for new channel. Get these from a steel supplier.

The hydraulic-jack capacity can range from 2 to 12 tons.

Cutting & Drilling—Don't wait to drill the holes until after welding or you'll never get it done. There

Trailer set up for hauling three motorcycles is light and tows easily behind compact car.

are about 28 1/2-in. holes to drill. Drilling individual parts is easier than drilling the welded assembly.

143

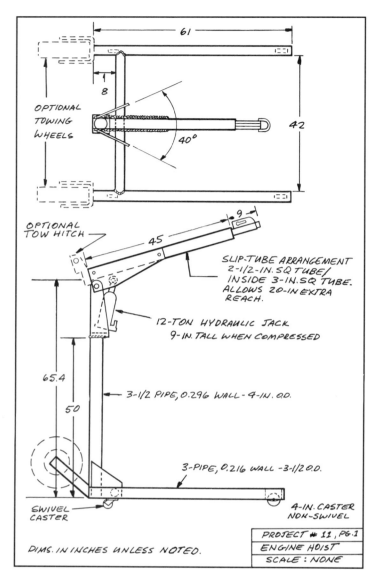

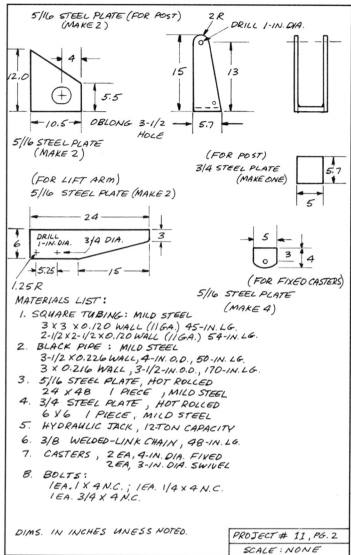

Press Dies & Collars—Make dies and most other accessories as needed. But, you will need various collars, sectioned from pipe, for mandrels.

PROJECT #11— ENGINE HOIST

Welding Process—Cutting torch and arc welder.

Materials—Most can be found at a steel-products warehouse. Be specific about pipe sizes and wall thicknesses if you expect this hoist to do its designed job. If the plans are followed precisely, the hoist will lift the largest passenger-car V8 engine and transmission

together. The 12-ton jack may be removed for use in the hydraulic floor press.

Welding—All welds must be *hot* for maximum penetration. Run two or three beads where possible. Weld all around and on both sides wherever possible.

Portability—Pneumatic boat-trailer tires can be added to the two lower beams and a 1-7/8-in. trailer hitch added to the rear of the arm to make the engine hoist towable.

PROJECT #12— TOW BAR FOR CARS & JEEPS

Welding Process—Cutting torch

and arc, TIG or wire-feed welding.

Materials—Buy only good-quality new materials. Two tow-bar capacities are described. The light tow bar will safely tow cars up to 3500-lb total weight; heavy bar up to 6000 lb. My light tow bar has pulled my Corvairs across the U.S. and a Scirocco race car up and down the west coast. If the tow bar is bolted snugly to the towed vehicle and careful towing habits are observed, cars will track well at all legal speeds.

Cutting & Fitting—A tow bar must be a perfect triangle to tow a car straight. Consequently, it's best to make a jig. Use a

Completed tow bar for Corvair race car hooked up and ready to go.

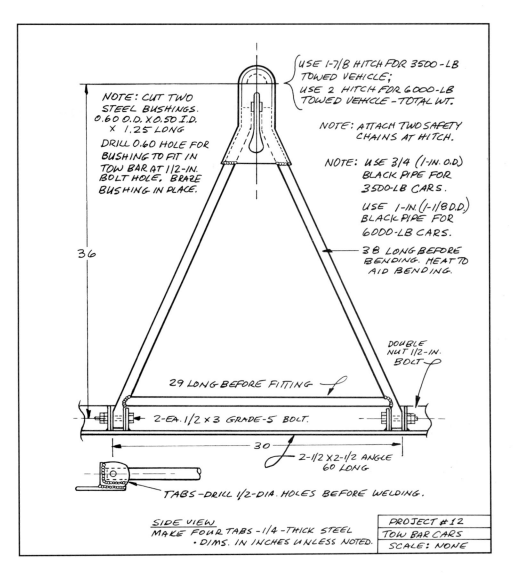

NOTE: CUT TWO STEEL BUSHINGS. 0.60 O.D. X 0.50 I.D. X 1.25 LONG DRILL 0.60 HOLE FOR BUSHING TO FIT IN TOW BAR AT 1/2-IN. BOLT HOLE. BRAZE BUSHING IN PLACE.

USE 1-7/8 HITCH FOR 3500-LB TOWED VEHICLE; USE 2 HITCH FOR 6000-LB TOWED VEHICLE – TOTAL WT.

NOTE: ATTACH TWO SAFETY CHAINS AT HITCH.

NOTE: USE 3/4 (1-IN. O.D.) BLACK PIPE FOR 3500-LB CARS.

USE 1-IN. (1-1/8 O.D.) BLACK PIPE FOR 6000-LB CARS.

36

3B LONG BEFORE BENDING. HEAT TO AID BENDING.

29 LONG BEFORE FITTING

DOUBLE NUT 1/2-IN. BOLT

2-EA. 1/2 X 3 GRADE-5 BOLT.

30

2-1/2 X 2-1/2 ANGLE 60 LONG

TABS – DRILL 1/2-DIA. HOLES BEFORE WELDING.

SIDE VIEW
MAKE FOUR TABS – 1/4-THICK STEEL
· DIMS. IN INCHES UNLESS NOTED.

| PROJECT #12 |
| TOW BAR CARS |
| SCALE: NONE |

1-7/8-in.-hitch ball as the center point to line up the hitch and lay out a triangle on a piece of 1/2-in. plywood. Use a 1/2-in. steel bar as the base of the triangle. Tack-weld the tow bar in the jig, then remove it for final welding.

Attaching to Tow Vehicle—On my Scirocco race car, I removed the front bumper guards, exposing a bolt hole under each guard. In each hole, I put a 3/8-in. Grade-5 bolt. I also drilled two more 3/8-in. holes on the bottom flange of the bumper for two more bolts—and the tow-bar angle. These four bolts and the two 1/2-in. hinge bolts were all double-nutted and checked every 100—200 miles of towing. The heavy-duty tow bar uses larger bolts, of course.

Tandem axle, four-wheel trailer will haul 5000-lb race car, four-wheel-drive vehicle, or even an airplane.

Loading ramps stow out of way under trailer.

145

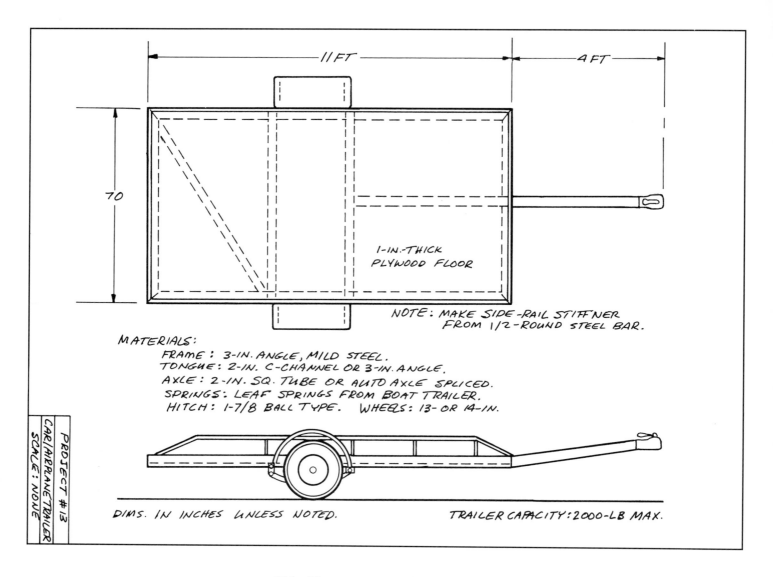

Dimensions and materials labels on drawing:

11 FT

4 FT

70

1-IN.-THICK PLYWOOD FLOOR

NOTE: MAKE SIDE-RAIL STIFFNER FROM 1/2-ROUND STEEL BAR.

MATERIALS:
FRAME: 3-IN. ANGLE, MILD STEEL.
TONGUE: 2-IN. C-CHANNEL OR 3-IN. ANGLE.
AXLE: 2-IN. SQ. TUBE OR AUTO AXLE SPLICED.
SPRINGS: LEAF SPRINGS FROM BOAT TRAILER.
HITCH: 1-7/8 BALL TYPE. WHEELS: 13- OR 14-IN.

DIMS. IN INCHES UNLESS NOTED.

TRAILER CAPACITY: 2000-LB MAX.

PROJECT #13
CAR/AIRPLANE TRAILER
SCALE: NONE

PROJECT #13—
TRAILERS FOR RACE CARS & AIRPLANES

Welding Process—Cutting torch, arc welding.

Materials—For the axle, I use the rear-axle assembly from a front-wheel-drive car. I just pick a wheel I like, and cut and splice or cut and narrow the axle to suit the desired dimensions. Springs can be purchased at a boat-trailer supply house or wheel manufacturer. *Never build a trailer without springs.* That $100 you might save by not buying springs could result in $1000 worth of damage to whatever you're hauling due to the rough ride. For a light trailer, try the rear axle from a 1980-or-later GM X-Body (Chevy Citation, Buick Skylark, etc.). For a heavier-duty axle, use a Cadillac Eldorado, Olds Toronado or 1979-and-later Buick Riviera rear axle.

Dimensions—These plans are for a trailer that will haul a 2000-lb car. Scale up the dimensions for a slightly larger car or down for hauling smaller items such as a home-built airplane. The trailer tongue can be lengthened at least 10 ft for airplane towing. Plans for a tandem-axle car trailer can be ordered from the J.C. Whitney Company.

PROJECT #14—
UTILITY TRAILER

Welding Process—Cutting torch, arc welding.

Materials—As in Project #13, I recommend using a rear-axle assembly from a front-wheel-drive compact car. If necessary, you can box-in the open U-channel or make a new beam axle from 2-1/2-in.-square tubing with a 0.250-in. wall thickness. Ordinarily, a utility trailer should be small and light. Therefore, the plans show a 6-ft-bed trailer. Improvise as you see fit.

PROJECT #15—
EXPERIMENTAL AIRPLANE

NOTE: Make sure you know what you're doing before you tackle this one. Start by contacting the Experimental Aircraft Association (EAA) and get the details on what's required of you.

Welding Process—Oxyacetylene, but TIG welding can be used if you're extremely good at it. Otherwise, use oxyacetylene.

Materials—Several manufacturers sell kits to build airplanes from 4130 steel tubing. Or you can get 4130 tubing and sheet from retailers and mail-order outlets:

Light utility trailer is good for hauling such items as snowmobiles and furniture.

Rear-axle assembly from front-wheel-drive Chevy Citation can be adapted to two-wheel trailer, Project #9. Shock-absorber, sway-bar and body-mount attachments will be cut off with a cutting torch. Axle will be attached to pair of leaf springs on trailer.

If front cockpit of partially-completed Christen Eagle biplane makes you want to build and fly your own airplane, order a set of plans and start welding. For information on welded-steel airplanes, contact: Experimental Aircraft Association, Wittman Airfield, Oshkosh, Wisconsin 54903-2591

WICKS AIRCRAFT SUPPLY
410 Pine Street
Highland, IL 62249
Phone (618) 654-7447

WAG-AERO
Box 181
Lyons, WI 53148
Phone (414) 763-9586

MONNETT EXPERIMENTAL
AIRCRAFT, INC.
Box 2984
Oshkosh, WI 54903

AIRCRAFT SPRUCE AND
SPECIALTY
Box 424
Fullerton, CA 92632
Phone (714) 870-7551

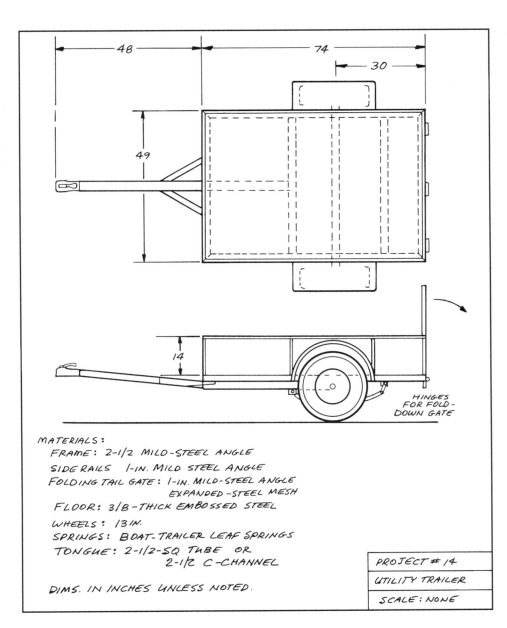

MATERIALS:
FRAME: 2-1/2 MILD-STEEL ANGLE
SIDE RAILS 1-IN. MILD STEEL ANGLE
FOLDING TAIL GATE: 1-IN. MILD-STEEL ANGLE
              EXPANDED-STEEL MESH
FLOOR: 3/8-THICK EMBOSSED STEEL
WHEELS: 13 IN.
SPRINGS: BOAT-TRAILER LEAF SPRINGS
TONGUE: 2-1/2-SQ TUBE OR
          2-1/2 C-CHANNEL

DIMS. IN INCHES UNLESS NOTED.

HINGES FOR FOLD-DOWN GATE

PROJECT # 14

UTILITY TRAILER

SCALE: NONE

Arts & Crafts Welding

Master blacksmith, Verne Monroe (right), and crew pose beside their forge while end of solid 8-in.-diameter steel round stock heats in furnace prior to forge welding. The year was 1915 at the Cameron Tool & Supply Company, Cameron, WV. Seventy years later, it's still in business doing custom forgings. Note leather aprons and size of sledge hammers. Verne Monroe was grandfather of co-author, Tom Monroe.

Put your imagination to work. Old sparkplugs were used to make airplanes. Tin strips were brazed to sparkplugs for wings. Butterfly wings were flame-cut from old chrome bicycle fender and brazed to horseshoe nail.

Welding can be a profitable and rewarding pursuit for the home hobbyist or part-timer. You could probably make some extra money welding broken bicycles, but you hate to take money for making kids happy. So, consider welding and brazing art. Most art shows feature some metal sculpture. The artist or craftsman simply uses his or her imagination to put objects of art together. You can start off with simple things and gradually progress into more-elaborate welding projects. Look at the nearby photos!

Welding decorative things can be immensely rewarding. If you admire the practical things you've welded, imagine how you'll feel after you do a piece of original weld art.

MATERIALS

In this chapter, I describe the materials and processes used to weld, cut, braze or solder the projects pictured. The materials can come from many sources, but if you plan to work with new materials, try the metals outlets in your area.

In all the countries I have worked in, I've found that "business pages" in the telephone book list the metal you are looking for. For instance, if you want sheet brass and sheet copper, look under **Copper** or **Brass** in your phone book. Most steel distributors and warehouses also carry other metals, such as aluminum, brass and copper sheet. Don't forget to check out model-train and model-airplane shops for brass tubing, brass and copper sheet, and even stainless tubing in very small, thin sizes. Jeweler's supply outlets may also have some of the more precious metals, such as silver.

You can also look for metal dealers in the classified advertising section of aviation magazines such as *Homebuilt Aircraft, Kit Airplanes, Aircraft Builder, Private Pilot* and *Sports Aviation*. Metals dealers listed in these magazines usually carry brass, stainless steel, galvanized sheet, aluminum and copper.

Try your local hardware store for sheet brass, copper tubing and other sheet metals. You'll also find things such as nails, nuts and bolts and many metal items that could be used in metal sculpture. It is easier to work with new metals because they'll be cleaner, but you can also try used metal items for your projects. For instance, an old chrome-plated bicycle fender was used to make the chrome-plated wings on the butterfly in the nearby photo. Many of the expensive metal sculptures you see in public places have been made from chrome-plated steel bumpers from auto-wrecking yards.

PROJECTS

I asked some metal-sculpture artists to describe the methods and materials used to produce the artwork shown in this chapter. In almost every case, the process was

Old paint cans were used to make candle holders. After cutting with small welding tip, cans were bent into spherical shape. Clean out all paint before applying flame!

Steel, brass and copper sheets were brazed together to make sailing ship for hanging on wall.

the same as for making a functional metal project. Drawings and sketches were made before starting the project. Jigs and fixtures were even used for some projects.

Candle Holder—The only arts and crafts project to *not* incorporate a drawing and materials list is the paint-can candle holder. The candle holders shown above were made from paint cans of varying sizes and were free-hand-cut and -shaped.

Old paint cans were first emptied of paint and then paint remover was used to strip all flammable materials from both inside and out. After the cans were water-rinsed and dried, the artist installed a very small tip—000—on an oxyacetylene cutting torch to free-hand-cut the decorative slots in the sides and lids of the paint cans. After the cans cooled, they were hand-formed to an oval-basket shape.

Sparkplug Airplane Mobile—A beginning weldor can complete this project in a day. Oxyacetylene cutting of steel sheet and brass brazing are the skills required. To duplicate this project, you will need two sparkplugs, a piece of copper-coated 1/8-in. steel welding

rod, and some sheet-metal pieces for the wings and tail. The artist who made the sparkplug airplanes used new sheet steel obtained from a metal-supply store, but you could use scraps of metal cut from old car fenders, metal serving carts, or even tin food cans.

Before starting, make a sketch of the finished project so you can cut the wings and tail pieces to correct proportions. Set the wings and sparkplug fuselage on a piece of firebrick or other non-flammable welding surface so they fit properly, then braze the parts together. Use the mechanical finger, page 52, to hold the parts for brazing. After the two airplanes have been brazed into finished assemblies, connect them into a mobile by brazing a 10-in.-long welding rod between them. Braze a carpet tack to the middle of the welding rod to act as a center bearing. The steel upright part of the base is made from a large nail, purchased at a hardware store. The nail head is center-punched to provide an indentation for the carpet-tack bearing to rest in. The large nail is then driven into a piece of driftwood to form the base.

Your imagination will provide ideas for similar metal sculpture. You might try to make a sculpture of the space shuttle or even a Fokker triplane from old sparkplugs.

The chrome-winged butterfly was made from a large horseshoe nail, pieces from an old bicycle fender, and small pieces of brazing rod. The entire assembly was brazed together and then glued to a small piece of decorative rock. A felt pad was then glued to the rockbase.

Brass Sailing Ship—New sheet brass was used to construct this wall decoration. The brass sheet was special-ordered from a metal-supply store. Because brass cannot be cut with an oxyacetylene torch, a small, high-pressure welding-tip flame was used to melt through the brass to obtain the sail patterns. Three-view sketches were made of the final assembly, and each sail was marked out on sheet brass to ensure proper shape and size.

The parts were assembled by silver soldering, using a solder with a very low melting temperature. Brazing rod would work, but the melting point of the thin brass sail and brazing rod would be the same, making it extremely difficult to braze without melting the sail. This sailing-ship wall decoration requires careful setup to assure that each sail is brazed into the correct position. Metal scraps can be used to block each sail in the correct place, but the mechanical finger really eases setup and positioning. Various sizes of brass brazing rod can be used to make the ship masts and rigging.

After the ship is completed, there will be a lot of brazing flux on each braze spot. The easy way to remove flux is to soak the entire sculpture in hot water for 10—15 minutes. Hot water will

SMITHING
by Kathleen Koopman

Before the advent of the modern-day metalworking, the forming of metal by heating, cutting, twisting, joining and filing was the blacksmith's job. Man's use of iron, bronze and other metals accelerated the growth of civilization. From the bronze age to the onset of the industrial revolution, it was the metalworker who was behind the success of farming and manufacturing enterprises. In more primitive societies, the metalworker was held in awe, even feared, because he worked with fire in semidarkness.

As civilization progressed, the metalworker—blacksmith—was respected in the community for his real and implied strength. He was relied upon to produce important household tools and hardware, farming implements and defense weaponry. These items were not only unique in their beauty, but practical and vital to the efficiency and prosperity of each particular society.

Smithing is a general term indicating the craft of working metals: the *whitesmith* worked with lead, the *blacksmith* with iron. A smith who made and fitted horseshoes was called a *farrier;* one who built wheels and wagons a *wheelwright.* There was also the *chainsmith,* the *cooper*—barrelmaker—and the *nailsmith,* each with his—her in the case of the nailsmith—own specialty. Smiths of all sorts were generally an independent lot. By necessity, they designed and made most of the tools and hardware used in their trade.

In the early days of the American colonies, metalworkers helped clear land, cultivate crops, transport supplies and products, make weapons for hunting and defense, and a whole variety of other improvements to the quality of life and growth of industries, such as mining. The blacksmith fashioned the ironwork that was so important to the growth of the individual and the country.

With the rise of the factory system and the industrialization of manufacturing at the end of the 19th century, the need for individually crafted, one-of-a-kind items dwindled. Many traditional blacksmiths joined ranks with factory workers, becoming machinists or mechanics. At about this time, gas- and electric-welding methods were developed, along with more powerful and effi-

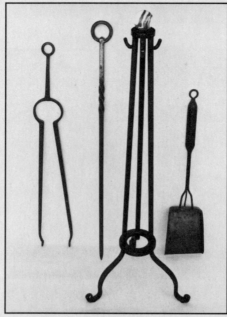

A lasting gift, Verne Monroe fabricated fireplace set for Christmas gift to one of his children in 1932. A combination of skills were required to make the pieces, one being forge welding. Ball on top of stand is old gearshift knob. Photo by Tom Monroe.

cient tools. Products that were previously a result of long hours at the forge and anvil could now be mass-produced and standardized as well. This enabled a greater part of a booming population to afford products such as the automobile, typewriter, camera and refrigerator.

Today, modern time-saving methods coupled with a back-to-basics movement in this country has contributed to interest in the age-old art of hammering iron—heated in a coal-fired forge—into shape on an anvil. Many backyard and homestead blacksmiths, using a combination of old and new methods, are springing up. They are supplying their own needs and their friends', and perhaps a small following looking for that one-of-a-kind item not available in any store.

In the past, the blacksmith provided practical, yet beautifully crafted implements. Today, the trend is toward personally designed pieces for the home and garden. Anyone can run down to the hardware store for a standardized nut or bolt. But, where can one go for a screen to fit a slightly irregular adobe fireplace built 100 years ago? Or a damper to build economy into that old flue? Today's blacksmith can do this and more.

Inspiration—The possibilities are endless for the home welder-smith

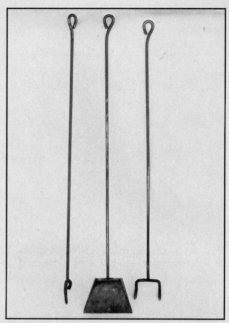

Fireplace set was done as a high-school metal-shop project by Verne Monroe's great-grandson, Jeff. Instead of forge welding, gas welding was used to join pieces. Photo by Tom Monroe.

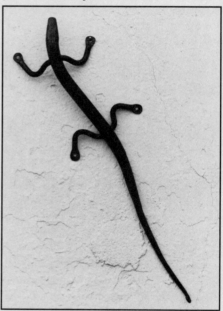

Lizard scurrying up garden wall is actually forged round stock heated, pulled and turned. Each pair of legs are one piece that passes through hole drilled in body. Holes were punched in feet to allow mounting. With little modification, lizard could be used as a door handle. Photo by Kathleen Koopman.

who wants to create beautiful, original items. The delicate balance between form and function provides a challenge any well-educated home craftsman should welcome. Inspiration is everywhere. Do you need to match an old hinge you found in the basement or at a swap meet to hang

Fireplace tong is made of round stock that was bent, coiled and flattened. Holes for rivet at pivot were pierced with punch. All operations require heating in furnace first. Photo by Kathleen Koopman.

Shovel was made using similar techniques as used for making tong. Shovel end was made from bent-and-folded sheet steel. Made by Richard Redman. Photo taken at the Country Emporium, Tucson, Arizona by Kathleen Koopman.

Rotating pot rack consists of tapered, curled and arc-welded round stock. Flat stock was for mounting bracket and circular section. Hooks can be moved or removed to accommodate different pots and utensils. Made by Richard Redman. Photo taken at the Country Emporium, Tucson, Arizona by Kathleen Koopman.

a door? How about some period hardware to adorn an antique you've just refinished? Could hanging potracks make your modern kitchen more convenient? Is there some quaint detail in your home you'd like to capitalize on? How about a hanging or mounted ornament to enliven your garden or entryway?

Many other sources of ideas are available. Your local library has many books and periodicals on blacksmithing, ironworking and ornamental metalwork. Most regions of the country have a rich architectural history that includes wrought ironwork. Most cities have museums and historical societies that offer a wealth of information, ideas and inspiration.

Use the materials you've discovered, plus photographs and sketches you've made of the things you like, as a starting point for a design. Work out small thumbnail sketches before attempting to work full-size. Using a broad-tipped marker—which will prevent you from being too detailed—sketch the design on large sheets of brown paper or corrugated cardboard. Don't rush the job. Take time to look at your pattern and make changes as needed before cutting the metal. Start with something small and simple, perhaps one that requires a basic structure with a decorative element added later.

Ironwork is an adaptable and flexible medium with endless design and creative possibilities for the home welder-smith. Many homeowners are involved in renovating or restoring old homes. This demands attention to detail, researching original plans and photographs.

By using the processes described in this book and a little creative thinking, many beautiful pieces can be made. Each of these items pictured were welded using a variety of old and new methods: coal-fired *forge welding,* gas welding and oxyacetylene welding.

The Process—As with any welding, the joints or seams must be correctly prepared. Because the pieces must overlap when they are forge-welded, they are cut at an angle, or one piece is V-notched and the other pointed so it will fit in the V-notch.

Forge welding is done by heating the workpieces in a low-sulfur coal- or coke-fired *forge*—furnace specially built for forging iron. Air can be blown up through the fire to increase heat. The pieces are heated to a *sweat heat*—temperature just below melting point. It takes a "good eye" to judge when this heat has been reached. The metal is virtually white hot.

While heating, the weld joints are coated with flux. It is gradually *spooned* on with a steel ladle. Pure borax was originally used to clean the metal, but has been replaced by a combination flux/filler material. The flux cleans and the filler fills voids in the metal. Flux is available from the Anti-Borax Company, Inc., 1502-1506 Wall Street, Ft. Wayne, IN 47804.

Once the metal reaches the sweat-heat point, the pieces are pulled out of the fire, overlapped or fitted together and hammered together on an anvil. A helper usually holds one piece while the blacksmith handles the other and does the hammering. Sparks fly everywhere as hot flux squirts from the weld joint. Correctly done, hammering fuses the parent and filler metal.

A New Renaissance—As a result of the combination of modern technology and the traditional methods of the past, today's metalworker is in a fortunate position. He can draw on the rich history of the craft, and reap the benefits of gas torches, arc-welding and electric power tools. Even the person who considers himself a "blacksmith" in the purest sense can turn to modern methods to expedite a task and *not* apologize.

Consider this: It was the smith with his forge and anvil that made this country strong. Perhaps a greater appreciation of the older methods of heating and pulling, twisting and hammering of iron in combination with a respect for the more modern methods will enable the skill of smithing to re-emerge in a Renaissance of handcrafts we are now experiencing.

Elaborate metal sculpture was made by brazing thousands of short pieces of welding rod together, then covering outer surface with melted brass rod. Assembly was chrome-plated afterward. Asking price for sculpture was $3000!

Sheet steel was used to fabricate ornate bird-claw lamp base.

F4-U Corsair model was made by fusion-welding hundreds of pieces of welding rod together. This combines exactness and patience of a model builder with skill of a weldor.

soften the flux so you can wash it off. After the brass sculpture has dried, spray it with clear lacquer to inhibit tarnishing. Otherwise, it'll turn green.

Longhorn Cow Skull—The elaborate metal cow skull was made by brazing together pieces of 1/16-in. steel welding rod. The horns were formed from sheet steel. One by one, thousands of short pieces of welding rod were fitted and brazed to form the basic structure.

After the basic structure was completed, the front of the skull was filled in with brazing rod melted and flowed into a smooth surface. Then, the brass surface was filed and polished to achieve a smooth finish.

The sheet-steel horns were rolled to form a funnel-shaped tube and then welded to eliminate the open seam. The horns were then heated red hot at each ripple and bent to the final shape. Finally, the horns were brazed to the skull. The entire assembly was polished in preparation for chrome plating.

The chrome-plating process included: copper-plating, polishing the copper, nickel-plating over the copper, then chrome-plating the assembly. This long and involved project gave fantastic results.

Bird-Claw Lamp Base—The accurately detailed lamp base shown above was made from 0.050-in. mild-steel sheet. The sheet was formed into a tapered tube and welded together to provide a welded-seam tube. More sheet was cut and fitted to make the leg, foot and claws. The pieces were then oxyacetylene-welded. The complex shapes of the claws and steel ball the claw is attached to were made by heating and forming the steel. Average time required to develop and weld a project such as this is about 1—2 weeks of full-time work.

Welding-Rod Airplane—This was made similar to the way you would build a balsa-wood-and-paper-covered airplane, except that steel welding rod was substituted for the balsawood sticks. Plans were drawn with soapstone on sheet-metal plate so the plane's wing ribs, fuselage sides, rudder and elevators could be flame-cut directly on the drawing. Ob-

viously, you could not assemble the welded model directly on a set of paper plans because you would incinerate the plans.

The airplane was oxyacetylene-welded, using a small torch. It can also be brazed if you prefer. One problem you'll encounter while brazing small parts together is affecting a second or third connection to a previously brazed joint. The heat required to add a piece often melts the previously brazed joint and the parts could come apart.

An advantage to fusion-welding a welding-rod airplane is that you are much less likely to melt the previous joint when adding a connection. I save my welding rod scraps by welding the short pieces together. Then I can use all of the rod. Try it—it's easy to do.

Like the proverbial boat in the bottle, a welding-rod airplane is a complex assembly that requires a lot of patience. It might take 40 hours to build just one model, but you'll surely impress your friends and yourself with the finished model.

Such is the nature of weld art.

Glossary of Terms

Every trade has its special terms and jargon. Learn the language of welding so the next time you go to the welding-supply store, you'll be able to ask intelligent questions. What follows is a handy weldor's glossary:

Air-Arc Gouging—An electric-arc process that cuts metal by melting it with a carbon or copper electrode, and simultaneously blows away the molten metal with a 100-psi air blast through the center of the electrode. It's a very noisy and messy process, but gets the job done cheaply where a lot of metal has to be removed.

Alloy—Basic metal modified by chemical compositions to improve its hardness or corrosion-resistant characteristics.

Aluminum Heat-Treating—Process by which aluminum is heated to 960—980F (516—527C), then quickly cooled, or *quenched*. The temperature is reached by placing the part in an oven or in a bath of liquid salts called *solution heat-treatment*. Quenching is accomplished by using cool air or water. Quick cooling is the secret to heat-treating.

Annealing—Opposite of hardening. It's done to remove hardness in certain metals where drilling or other machining is desired. The metal is usually heated to about the same temperature as for heat-treating, but then allowed to *cool slowly*. Aluminum and steel may be annealed. Usually, the part can be heat-treated again after annealing.

Arc Blow—Deflection of the arc from its normal path by magnetic forces, usually associated with d-c welding.

Arc Welding—A welding process that fuses metal by heating it with an electric arc and simultaneously depositing the electrode in the molten puddle.

Backfire—Momentary, loud *pop* at the oxyacetylene torch tip. It is caused by the flame backing up into, or combustion occurring inside, the tip. It's usually a result of the weldor trying to get more heat from a torch with low gas pressure by holding it too close to the work and overheating the tip. This occurs more readily in a large torch with a *rosebud* tip because of low gas pressure. Backfire can be dangerous and should not be allowed to continue.

Backhand Welding—Like walking backward, it is welding backward. The weldor points the torch at the already welded seam, away from the unwelded seam. I doubt the need for this procedure, but it could be useful to avoid burning through very thin metal. The added mass of weld bead could help absorb the extra heat.

Backing Ring—Metal ring placed inside the seam of pipe being *butt-welded*—welded end to end. The ring provides for full weld penetration and 100% strength in butt-welded pipe seams. It is usually tapered for smooth flow of liquids, steam or gas once the weld is completed. The ring is then left inside the pipe.

Backing Strip—Metal strip that serves the same purpose as a backing ring—to ensure 100% strength and weld penetration.

Bare Electrode—A consumable, bare electrode used in arc welding with no flux coating.

Base Metal—Prime metal to be welded, brazed or cut. In auto bodywork, the car's steel fender is the base metal and the welding rod the filler metal.

Bead or Weld Bead—Result of fusing together a seam in two or more pieces of metal with welding rod. Usually, the bead is thicker than the base metal.

Bevel—Preparatory step prior to welding, whereby the edges of the base metal are angle-filed or ground to better accept the filler metal. When welding thick metal, the bevel forms a V-groove that promotes better weld penetration.

Braze—Non-fusion weld produced by heating a base metal above 800F (427C) and using a non-ferrous filler metal. The liquid (above 800F) filler metal flows between closely fitted surfaces of the metal joint by capillary action. The base metal is not melted in braze welding.

Butt Joint—Joint between two pieces of metal lying flat, end to end.

Capillary Action—Action whereby the surface of a liquid—including metal heated to a liquid state—is raised, lowered or otherwise attracted to fixed molecules nearby. A plumber *sweats* together copper pipe by using liquid brass and other liquid solder to flow into tight-fitting places by capillary action. Dictionary defines it as "the force of adhesion between a solid and a liquid."

Carburizing—1. A heat-treating process that hardens iron-based alloys by diffusing carbon into the metal. The metal is heated for several hours while in contact with carbon, then quenched; 2. In gas-welding, an acetylene-rich flame that coats the metal with black *soot*.

Corrosion—Gradual chemical attack on metal by moisture, the atmosphere or other agents. This includes rust on iron or steel, oxidation on aluminum, and acid pitting and etching on stainless steel. Corrosion is the biggest long-term problem for fabricated metal construction. Corrosion prevention can be accomplished by painting, plating, oiling or any other coating that keeps oxygen away from the base metal.

Cover Glass—Clear glass used in goggles and welding helmets to protect the more-expensive colored lens from weld spatter.

Covered Electrode—Arc-welding electrode, used as a filler metal, that is covered with flux to protect the molten weld puddle from the atmosphere until the puddle solidifies. Commonly called *stick electrode*.

Deposit—Filler metal added during the welding operation.

Depth of Fusion—Depth that fused filler material extends into the base metal.

Duty Cycle—Seldom-understood term that applies to electric-arc welders, not gas welders. The duty cycle is a ten-minute period. If an arc welder has a 100% duty cycle, it can be used 10 minutes out of every 10 minutes or 100% of the time. If a welder has a 20% duty cycle, it can be used two minutes and must cool off for eight minutes out of each 10 minutes. Usually, a small arc welder will have a 90% duty cycle at the lower amp settings, tapering off to 5% at the highest settings.

Dye-Penetrant Testing—An inexpensive process to check welds for cracks and other defects. The process consists of three chemicals in solution: a cleaner, a spray-on or brush-on penetrating red dye, and a white developer solution. After the area is cleaned and the red dye allowed to soak in a few minutes, the developer is sprayed on. Defects show red and smooth areas appear white. After inspection, the cleaning solution can be used to remove the dye.

Field Weld—Weld done at the site or in the field rather than in a welding shop.

Filler Metal—Welding rod or other metal added to the seam to assure a maximum-thickness weld bead.

Fillet Weld—Weld deposit of filler metal approximately triangular in shape. Usually made when welding a T-joint or 90° intersection.

Flash Burn—Burn caused by ultraviolet-light radiation from the arc in arc welding. Usually painful and more severe than sunburn, especially when the eyes are flash burned.

Flashback—Burning of mixed gases *inside* the torch body or hoses. Usually accompanied by a loud hiss or squeal. *Must not be allowed to continue!* Shut off *oxygen* immediately if flashback occurs, then shut off acetylene. Turn off oxygen first because it supports combustion. Use in-line arrestors to prevent flashback. If flashback occurs, do not light up the torch again until you find the cause and eliminate it.

Flux—Chemical powder or paste that cleans the base metal and protects it from atmospheric contamination during soldering or brazing. Flux consists of chemicals and minerals that properly clean and protect each type of metal. Therefore, each type of metal joining requires a *specific* formulation of flux. Flux is *not* used in fusion welding, except as a coating over arc-welding rod or in submerged-arc welding.

Fusion Welding—The only true kind of welding. The metal pieces to be welded are heated to a liquid state along the weld seam, and, usually filler metal of the same or similar type is added to the molten puddle and allowed to cool, forming one continuous piece of metal. Thus, the weld should be stronger than the added filler metal.

Gas Welding—Also known as *oxyacetylene welding*. Common term to describe welding accomplished by burning oxygen and acetylene to make a 6300F (3482C) flame.

Gas Metal-Arc Welding—Or GMAW, a process in which an inert gas such as argon, helium or carbon dioxide is fed into the weld to shield the molten filler metal. This displaces atmospheric air and inhibits oxygen from combining with the molten metal and forming oxides and other impurities that would weaken the weld. Also known as metal inert-gas—MIG—welding, but commonly called wire-feed welding.

Gas Tungsten-Arc Welding—Or GTAW, another type of inert-gas arc welding with tungsten as the electrode material. Tungsten is used because it will not melt at welding temperatures. The arc is similar to the heat from an oxyacetylene torch except that it can be concentrated in a much smaller space. Helium or argon gas is used to shield the weld puddle; argon is preferred. The filler metal is uncoated. This process can be used to weld steel, stainless steel, titanium, aluminum, magnesium and several other metals. Also known as tungsten inert-gas—TIG—welding and commonly known as Heliarc welding. Heliarc is a registered trademark of Linde.

Hard Facing—Applying a very hard metal face to a softer metal to improve wear characteristics, such as on a bulldozer blade. The welding rod itself is the hard metal.

Hardness Testing—There are three types of tests: Rockwell, Brinell and Shore. Rockwell Hardness is a two-stage ball-impression test. Brinell Hardness is a one-stage ball-indentation test. Shore Hardness is a drop test for indenting metal. All three hardness tests use the principle that the harder the metal is, the less likely it is to be dented by a given force. Obviously, all three methods require calibrated special equipment, and a trained operator to analyze the results.

Heat-Affected Zone—Portion of the base metal that has not melted, but that has become discolored or *blued* by the heat from welding or cutting. Usually, metal strength is changed in the heat-affected zone. The heat-affected zone is usually detectable by eye.

Heat Sink—A mass of metal, water-soaked rag or other heat-absorbing material, placed so it absorbs heat, thus preventing overheating of a component or area. The use of a heat sink in welding can prevent or limit burn-through or warpage.

Heat-Treating—A process that adds strength and brittleness to metal. Almost all metals have a critical temperature at which their grain structure changes. This involves controlled heating and cooling of the metal to achieve the desired change in crystalline structure. Not all metals can be heat-treated.

Heliarc—Trademark of Linde. See *TIG* and *Gas Tungsten Arc Welding.*

Horizontal Position—Weld seam is horizontal.

Interpass Temperature—When several passes or beads are made in welding a joint, the lowest temperature of the weld bead before the next pass is started. Keep it as low as possible when welding cast iron to prevent cracking of the weld.

Joint—Junction where two or more metal pieces are joined by welding, brazing or soldering.

Kerf—Width of the cut, in oxyacetylene or plasma-arc cutting.

Keyholing—Usually occurs when butt-welding two very thin pieces of aluminum. A small hole melts all the way through but is filled with filler rod.

Magnaflux—A magic word in the welding or metallurgy business. An inspection process used with magnetic (ferrous) materials to detect cracks or other flaws. A fine iron powder is sprayed over the inspection area, then a strong magnetic field is induced electrically to cause any crack or defect to show as a separation of the iron particles. In most cases, this process causes the part to become magnetized, requiring demagnetization after the inspection is complete. Industrial Magnaflux equipment is similar in size and cost to a large arc welder.

Melting Point—Temperature at which a metal melts and becomes liquid. The melting point of frozen water (ice) is 32F (0C). Mild steel's melting point is 2700F (1482C). See the melting-point chart, page 7.

MIG Welding—Metal inert-gas, or wire-feed, welding. The "M" for metal is a spool of wire fed through a shield of inert gas, usually 25% helium and 75% argon. It produces a result similar to tungsten inert-gas—TIG—welding. In the case of MIG welding, the filler rod or wire becomes the electrode and melts. Therefore, it must be continuously *fed* to the weld puddle. Stainless steel, steel, aluminum and other metals that can be fusion-welded can also be MIG-welded.

Neutral Flame—An oxyacetylene flame with equal amounts of oxygen and acetylene.

Nitriding—A surface-hardening process for certain steels, whereby nitrogen is introduced in contact with anhydrous ammonia gas in the 935—1000F (502—538C) range. Quenching is not required. It's accomplished in similar manner to carburizing.

Normalizing—This process is usually used on high-strength steels such as 4130 to remove strains in fabricated parts or in material intended for bending or machining. The metal is heated to a point above the critical transformation temperature and then allowed to cool in still air at room temperature—with no drafts.

Overhead Position—Welding position in which the seam is welded from the underside of the joint.

Oxidize—1. Effect of applying excess oxygen, causing metal to vaporize during welding, as in an *oxidizing* flame; 2. Slow chemical process whereby oxygen and water combine to attack ferrous metals, resulting in rust and corrosion.

Oxidizing Flame—Gas-welding flame with excess oxygen.

Oxygen Cutting—Special cutting process that cuts metals by the chemical reaction of oxygen and the base metal at elevated temperatures. No acetylene is used except to start the oxidizing process, then only oxygen is used.

Peening—Working metal with a small pointed hammer or spraying with small steel shot. Peening is usually done to improve the surface strength of the metal by putting it in compression. This prevents cracks from starting at the surface.

Penetration—Depth of weld metal from molten puddle into the base metal. Ideal penetration is 15% to 110%. Less penetration makes a weak weld. At more than 100%, the weld bead is thicker than the base metal.

Pickling—Just like putting cucumbers in solution, you put metal in a dilute acid or other chemical to clean oil, scale and other unwanted matter from its surface. Usually, a corrosion inhibitor such as wax, is applied to the surface after pickling to prevent corrosion until the metal is painted or plated.

Plating—Outer coating of chromium, copper, nickel, zinc, cadmium or other heavy metal to enhance appearance or inhibit corrosion of parent metal. Usually, plating is accomplished by immersion in an acid solution with cathode and anode electric current, causing the plating material to deposit on the parent metal. Most ferrous metals can be plated.

Plug Weld—Also called *rosette weld.* A circular weld made through a hole in one piece of tubing or blind channel, connecting another piece slipped inside.

Polarity—In direct-current arc welding, TIG welding and wire-feed welding, how current flows; either positive to negative, or negative to positive. *A-c* welding has no polarity because it switches between positive and negative polarity and back again 60 times per second. *Heat is about equal at the electrode and workpiece. D-c, straight polarity* (DCSP) is also called *negative polarity.* In DCSP, the workpiece is positive and the electrode negative. *Heat is greater at the workpiece than at the electrode. D-c, reverse polarity* (DCSP) is also called *positive polarity.* In DCRP, the workpiece is negative and the electrode positive. *Heat is greater at the electrode than at the workpiece and penetration is shallow.*

Postheating—Heating a weld after it is completed, usually for stress relief.

Preheating—Heating the weld area beforehand to avoid thermal shock and thermal stresses. Metal is more sensitive to temperature than you would think. For instance, a steel pipe welded at 32F (0C) will be more brittle than one heated to 90F (32C), or even to 300F (149C). The larger the mass to be welded, the more it needs preheating.

Puddle—Liquid area of the weld where heat is being applied either by flame or electric arc. *This is the most important part of welding!* If the puddle is properly controlled, the weld will automatically be good.

Reducing Flame—In oxyacetylene welding, a flame with excess acetylene, imparting excess carbon to the weld. Also called a *carburizing flame.*

Residual Stress—Stress remaining in a structure after the weld joint cools. Metal wants to warp when it is heated. If a very strong jig or a triangulated structure prevents warping, stresses remain in the area near the weld.

Root Opening—Distance between two pieces of metal to be joined.

Root of Weld—Point of weld farthest from heat source. Intersection point between bottom of weld and base-metal surface.

Rosette Weld—See *Plug Weld.*

Sandblasting—Fast, easy method of cleaning certain metals before welding and painting. A high-velocity air blast, carrying sand, is directed at the metal and the particles of sand abrade its surface. Obviously, this cleaning process should be used with caution to protect eyes and lungs. It leaves a rough surface and cleanup is messy.

Seam Welding—A form of spot welding in which two pieces of sheet metal are resistance-welded in a continuous seam.

Shielded-Metal Arc Welding—SMAW, also commonly known as *stick welding* or as the layman knows it—simply *arc welding.* Shielding is the flux coating on the metal rod.

Shop Weld—To prefabricate or weld subassemblies in a shop or controlled environment before taking them on-site for final assembly. Often used in large welding projects such as oil-drilling sites and nuclear power plants.

Slag—Impurities resulting from heating metal and boiling off dirt and scale present in most open-air welding. Slag is found at the kerf from oxyacetylene-torch cutting. Slag will also be found in the hardened flux on top of an arc-welded bead.

Soldering—Metal-joining process similar to brazing. Metal pieces are joined with molten solder, without melting the base metal. Solder is drawn into the joint by capillary action. As it cools, it sticks to the base metal. If two pieces of lead solder were joined by melting them together, technically that could be called welding. But I would just call it melting lead!

Spatter—Small, unsightly droplets of metal that deposit alongside the weld bead. Especially common in arc welding with E-6011 rod.

Spot Welding—A production-welding method to join sheet metal. Electrical resistance heating and clamping pressure are used to fuse panels together with a series of small "spots." Filler metal is *not* used. Also called *resistance welding*.

Steel Heat-Treating—Process of heating and rapidly cooling steel in the solid state to obtain certain desired properties: workability, microstructure, corrosion resistance, and so on. Depending on its mass and alloy type, the steel is oven-heated to 1475—1650F (802—900C), then quenched by dipping it in water or oil.

Stickout—Length of electrode (tungsten or wire) that sticks out past the gas lens, cup or gun.

Stitch Weld—Tack-welding technique with short weld beads about 3/4-in. long, spaced by equally long gaps with no welding. Used where a solid weld bead would be too costly and time-consuming, and where maximum strength is not required.

Stress Cracking—Metal cracking at the weld due to temperature changes or molecular changes. Overheated welds are more prone to stress cracking than underheated welds.

Stress-Relief Heat-Treating—When a complicated, rigidly braced structure such as an airplane engine mount or race-car suspension member is welded, stresses remaining in the metal will cause premature fatigue cracking unless they are relieved. Stress relieving is accomplished by heating part or all of the structure to about two-thirds of the melting point and then cooling it slowly. This allows the molecules in the structure to relax and stay relaxed.

Stringer Bead—A straight weld bead made without oscillation.

Submerged-Arc Welding—SAW or sub-arc welding. A process in which the electric arc is submerged in powder flux, thereby protecting the weld from atmospheric

contamination. The system is usually automatic-feed and -travel, and a base-metal rod is used. It's used where high accuracy and weld quality are desired.

Sugar—Crystallization in a weld. It usually occurs when welding stainless steel if the back side of the weld seam is not protected by an inert gas such as argon. Sugar has no strength and should not be allowed in a weld.

Tempering—When metal has been hardened by heat-treating, it usually becomes brittle. In order to relieve the internal strains, the metal is usually reheated to about one-quarter or one-half the temperature originally used in heat-treating.

Thoriated Tungsten—Tungsten electrode with 1—2% thorium added to provide a more stable arc. Thoriated tungsten is used to weld steel. However, pure tungsten must be used to weld aluminum and magnesium.

TIG—Tungsten inert-gas welding. Often called Heliarc, a Linde trade name, because helium was first used as an inert gas for this welding process. Argon and other gas mixtures are now used. Electrode is tungsten because it doesn't melt at welding temperatures. The inert gas shields the weld from atmospheric impurities, providing a high-quality weld. Filler rod is hand fed to the weld. This process is also called gas tungsten-arc welding.

Tungsten Electrode—A non-consumable electrode used in TIG welding. The melting point of tungsten is 5432F (3000C).

Ultrasonic Testing—Process to test metal parts for defects. High-frequency sound waves are directed at the part, and their reflections are picked up by a receiver. Cracks and flaws inside the metal are detected by discontinuities in the

return sound. Ultrasonic test equipment is expensive and requires trained personnel to operate and analyze the results.

Vertical Position—Type of weld in which the metal to be welded is vertical and the weld bead progresses upward or downward.

Weave Bead—Weld bead made with a transverse oscillation such as a figure-8, or a "Z" motion while moving forward along the seam. This bead deposits more filler metal and ties the two pieces together more effectively than a straight bead, but also provides the possibility of including slag or flux in the weld, thereby contaminating it.

Weld—Local melting together and fusing of metal produced by heating the base metal and, in most cases, applying filler rod to the molten puddle. The filler rod usually has a melting point approximately the same as the base metal, but above 800F (427C).

Weldability—Capacity of specific metals to be welded and to perform satisfactorily for the intended service. *Not all* metals are weldable. See Chapter 1 for what can and cannot be welded.

Welder—Piece of equipment used for welding.

Weldor—Person who performs the welding operation.

X-Ray—Inspection for stress cracks, internal corrosion or other defects. An actual X-ray picture is taken of the part. The operator must be specially trained because of the radioactive materials involved. This process is used to inspect welded seams on high-pressure nuclear power-plant piping, airplane parts and other situations in which any defect would cause an expensive or potentially dangerous problem.

TERMS FOR WELDING DEFECTS

Arc Strike—Unintentional arc start outside of the weld bead. Usually more of a problem in TIG welding in which strict quality-control standards are observed.

Cold Weld—Poor penetration of the weld bead, usually less than 5% of the bead thickness.

Crater—In arc welding and TIG welding, a depression at the end of the weld bead caused by stopping the weld with too much heat applied.

Crater Crack—Crack in the crater at the end of the TIG-weld bead caused by stopping the weld with too much heat applied and withdrawing the shielding gas before the weld solidifies.

Drop Through—Filler material that sags through on the underside of the weld, caused by either too much heat or poor joint fit.

Discontinuity—An interruption in the basic weld bead, usually excess filler material, but not necessarily a defect.

Inadequate Penetration—Depth of filler metal is less than 15% of the weld-bead thickness.

Porosity—Usually, gas pockets caused by the wrong weld temperature or a dirty, contaminated weld. Most porosity is caused by getting the weld bead too hot.

Slag Inclusion—Dirty weld due to flux trapped in the weld bead or scale or dirt from the base metal or welding rod.

Undercut—Cutting away of the base metal by improper application of temperature. In the simplest terms, the weldor just pointed the heat at the weld bead and capillary action pulled the molten puddle away from the cooler base metal to the hotter molten puddle. Undercut can be avoided by more careful attention to temperature control.

Suppliers List

Suppliers are listed two different ways. First, they are listed by company name in alphabetical order. Then, the products they offer are listed alphabetically. If you are looking for a particular product, look first in the product list. The number next to each product refers to a company(ies) in the alphabetical suppliers list.

This a living document. New companies are formed and others go out of business, as time goes by. Some move to new headquarters without informing us. So, although we try, don't consider this list to be the final word. Check the advertisements in the latest trade and racing publications for welding-related products. The following suppliers list is merely a guide to get you started on your shopping task.

LIST OF SUPPLIERS

1. **AEG Power Tool Corp.**
   1 Winnenden Rd.
   Norwich, CT 06360

2. **Acco Industries Inc.**
   101 Oakview Drive
   Trumbull, CT 06611

3. **Aeroquip Corp.**
   300 S. East Ave.
   Jackson, MI 49203

4. **Airco Welding Products**
   575 Mountain Ave.
   Murray Hill, NJ 07974

5. **Aladdin Welding Products Inc.**
   1300 Burton St.
   Grand Rapids, MI 49507

6. **Alloy Rod**
   Allegheny International Co.
   Two Oliver Plaza
   P.O. Box 456
   Pittsburg, PA 15230

7. **Allweld Alloys**
   2027 Laura Ave.
   Huntington Park, CA 90255

7A. **All State Welding Products**
   5112 Allendale Lane
   Taneytown, MD 21787

8. **Aluminum Co. of America**
   1204 Alcoa Building
   Pittsburgh, PA 15219

9. **American Industrial Equipment Corp.**
   116 49th St.
   Union City, NJ 07087

9A. **American Optical Corp.**
   Safety Products Division
   14 Mechanics St.
   Southbridge, MA 01550

10. **Amsco Division**
   Abex Corp.
   389 E 14th
   Chicago Heights, IL 60411

11. **Anchor Clothing**
   Nasco Inc.
   4300 N. Florence St.
   Birmingham, AL 35217

12. **Anti-Borax Compound Co., Inc.**
   1502-1506 Wall St.
   Fort Wayne, IN 47084

13. **Arcair Co.**
   N. Memorial Drive
   Lancaster, OH 43130

14. **Arcos Corp.**
   1500 S. 50th St.
   Philadelphia, PA 19143

15. **Bausch and Lomb**
   Rochester, NY

16. **Cayuga Machine & Fabricating Co., Inc.**
   200 Gould Ave.
   Depew, NY 14043

17. **Ceramic Nozzles Inc.**
   201 Park Ave.
   Hicksville, NY 11802

18. **Chemetron Welding Products**
   Chemetron Corp.
   111 E. Wacker Drive
   Chicago, IL 60601

19. **Conley and Kleppen Enterprises, Inc.**
   3501 C St. NE
   Auburn, WA 98002

20. **Controlled Systems**
   2862 Kew Drive
   Windsor, Ontario, Canada
   N8T 3C6

21. **Cooperheat**
   206 Riverside Drive
   Newport Beach, CA 92663

22. **Cronatron Welding Systems, Inc.**
   141 Algonquin Parkway
   Whippany, NJ 07981

23. **Dockson Corp.**
   3839 Wabash Ave.
   Detroit, MI 48208

24. **ESAB Inc.**
   7255 Standard Drive
   Hanover, MD 21076

25. **Engelhard Industries Div.**
   Rt. 152
   Plainville, MA 02762

26. **Eutectic Corp.**
   40-40 172nd St.
   Flushing, NY 11358

27. **Flood Safety Products**
   1537 Walnut
   Kansas City, MO 64108

28. **Forney Industries Inc.**
   1830 Laporte Ave.
   P.O. Box 563
   Fort Collins, CO 80521

29. **GTE/Sylvania Inc.**
   Chemical and Metallurgical Division
   Hawes St.
   Towanda, PA 18848

30. **General Scientific Equipment Co.**
   Philadelphia, PA

31. **General Fire Extinguisher Corp.**
   1685 Shermer Rd.
   Northbrook, IL 60062

32. **Gullco International Inc.**
   4643 Giles Rd.
   Cleveland, OH 44135

33. **Harris Calorific Division**
   Emerson Electric Co.
   400 Clark St.
   Elyria, OH 44036

34. **Handy and Harman**
   850 Third Ave.
   New York, NY 10022

35. **Hobart Bros. Company**
   600 W. Main St.
   Troy, OH 45373

36. **Hoffman-Chicago Inc.**
   4635 S. Harlem Ave.
   Berwyn, IL 60402

37. **Huntsman Products Division**
   Kedman Co.
   P.O. Box 25667
   Salt Lake City, UT
   84125-0667

38. **J.W. Harris Co., Inc.**
   10930 Deerfield Rd.
   Cincinnati, OH 45242

39. **Jackson Products**
   Airco Inc.
   5523 E. Nine Mile Rd.
   Warren, MI 48091

40. **K. Woods, Inc.**
   Rt. 2, Box 230
   Washington, MO 63090

41. **Kaiser Aluminum and Chemical Corp.**
   300 Lakeside Drive
   Kaiser Center
   Oakland, CA 94643

42. **Kester Solder Division**
   Litton Systems, Inc.
   Chicago, IL

43. **Lenco, Inc.**
   319 W. Main St.
   Jackson, MO 63755

44. **Linde Welding & Cutting Systems**
   P.O. Box 6000
   Florence, SC 29501

45. **Magnaflux Corporation**
   7300 W. Lawrence Ave.
   Chicago, IL 60656

45A. **Maitlen & Benson, Inc.**
   P.O. Box 4146
   Long Beach, CA 90804

46. **Mid-States Steel and Wire**
   Division of Keystone
   Consolidated Industries
   510 S. Oak
   Crawfordsville, IN 47933

47. **Miller Electric Mfg. Co.**
   718 S. Bounds St.
   Appleton, WI 54911

48. **Mine Safety Appliances Co.**
   600 Penn Center Blvd.
   Pittsburgh, PA 15235

49. **Nemco Welding Alloys Div.**
   New England Metal
   P.O. Box 6647
   Providence, RI 02940

50. **Norris Industries**
   P.O. Box 7486
   Longview, TX 75607

51. **Norton Co.**
   1 New Bond St.
   Worcester, MA 01606

52. **O.K.I. Supply Co.**
   7584 Reinhold Drive
   Cincinnati, OH 45237

53. **Omega Engineering, Inc.**
   Box 4047, Springdale Station
   Stamford, CT 06907

53A. **Page-Wilson Corp.**
   Page Welding Division
   205 Clay St.
   Bowling Green, KY 42101

54. **Palmgren Steel Products**
   Division of Chicago Tool & Engineering Co.
   8383 S. Chicago Ave.
   Chicago, IL 60617

55. **Phoenix Products Co.**
   4715 N. 27th St.
   Milwaukee, WI 53209

56. **Rexarc, Inc.**
   P.O. Box 47
   West Alexandria, OH 45381

57. **Rockwell International**
   Power Tool Div.
   400 N. Lexington Ave.
   Pittsburgh, PA 15208

58. **Ryobi America Corp.**
   1158 Tower Lane
   Bensenville, IL 60106

59. **Sciaky Bros. Inc.**
   4915 W. 67th St.
   Chicago, IL 60638

60. **Shannon Marketing**
   P.O. Box 378
   24403 E. Welches Rd.
   Welches, OR 97067

61. **Sioux Tools, Inc.**
   2901 Floyd Blvd.
   Sioux City, IA 51102

62. **Smith Equipment Division**
   Tescom Corp.
   2600 Niagra Lane N.
   Minneapolis, MN 55441

63. **Solar Flux**
Golden Empire Corp.
3121 W. 139th St.
Palos Verdes, CA 90274

64. **Stoody Co.**
16425 Gale Ave.
Industry, CA 91745

65. **Techalloy Co., Inc.**
Rt. 113
Rahns, PA 19426

66. **Tempil Division**
Big Three Industries, Inc.
Hamilton Blvd.
South Plainfield, NJ 07080

67. **The Ansul Co.**
Marienette, WI

68. **The De Vilbus Co.**
837 Airport Blvd.
Ann Arbor, MI 48104

68A. **The Eastwood Co.**
P.O. Box 296
Malvern, PA 19355

69. **The Lincoln Electric Co.**
22801 St. Clair Ave.
Cleveland, OH 44117

70. **The Ridge Tool Co.**
400 Clark St.
Elyria, OH 44035

71. **Thermacote-Welco Co.**
32311 Stephenson Hwy.
Madison Heights, MI 48071

72. **Thermal Dynamics Corp.**
Industrial Park 2
West Lebanon, NH 03784

72A. **Titanium Wire Corp.**
110 Lehigh Drive
Fairfield, NJ 07006

73. **TWECO Products, Inc.**
P.O. Box 12250
Wichita, KS 67277

74. **U.S. Cylinders, Inc.**
100 Industrial Park
Citronelle, AL 36522

75. **Unibraze Corp.**
7502 W. State Rt. 41
Covington, OH 45318

76. **Uni-Nor Welding Products, Inc.**
310 Port Jersey Blvd.
Jersey City, NJ 07305

77. **Victor Equipment Co.**
Airport Rd.
P.O. Drawer 1007
Denton, TX 76201

78. **Wall Colmonoy Corp.**
19345 John R
Detroit, MI 48203

79. **Walter Kidde & Co., Inc.**
675 Main St.
Belleville, NJ 07109

80. **Weld-Aid Products**
15600 Woodrow Wilson
Detroit, MI 48238

81. **Weldcraft Products Inc.**
119 E. Graham Place
Burbank, CA 91502

82. **Westinghouse Electric Corp.**
Industry Automation Division
400 Media Drive
Pittsburg, PA 15205

83. **Williams Low-Buck Tools**
4175 California Ave.
Norco, CA 91760

84. **Wilson Sales Company, Inc.**
2112 Santa Anita Ave.
P.O. Box 3155
South El Monte, CA 91733

PRODUCT INDEX

Numbers refer to those shown in alphabetical suppliers list.

Anti-Spatter Compounds 26, 35, 39, 44, 75
Arc Stabilizers, High-Frequency 4, 21, 24
Arc Welders, A-C/D-C 4, 13, 24, 26, 28, 44, 47, 59, 69, 82
Arc Welders, A-C Transformer 4, 24, 47, 69
Arc Welders, D-C Motor Generator 4, 24, 28, 69
Arc Welders, Engine-Driven 4, 24, 28, 35, 47, 69
Brazing Alloys 4, 8, 18, 26, 38, 44, 71, 75
Brazing Fluxes 18, 23, 36, 44, 75, 78
Carbon-Arc Torches 39, 69
Clothing, Weldor's Protective 9A, 11, 15, 27, 35, 43, 75, 84
Cut-Off Machines, Abrasives 1, 57, 70

Cylinders, High-Pressure Industrial & CO_2 4, 44, 50, 74
Cylinders, Low-Pressure 4, 44, 74
Cylinder Carts 4, 62
Electrode & Flux Ovens 21, 32, 55
Electrodes, Coiled & Spooled 4, 14, 18, 24, 35, 41, 44, 46, 49, 64, 65, 69, 72A, 75
Electrodes, Coiled & Spooled, Tubular, Flux-Cored 18, 35, 44, 64, 69, 75
Electrodes, Stick 6, 18, 24, 35, 69, 75, 76
Electrodes, Tungsten 29, 35, 36, 75, 81
Eye Drops 30, 48
Fire Extinguishers 31, 67, 79
Flashback Arrestors 22, 33, 56, 75, 77
Flowmeters 4, 33, 44, 62, 75, 77
Fluxes 12, 14, 24, 35, 44, 52, 63, 69, 75
Gages 4, 23, 75
Gas Hose 3, 23, 33, 44
Grinders, Saws 2, 61

Helmets, Arc-Welding 23, 27, 37, 39, 44, 69, 75
Inspection Equipment 45, 58
MIG-Welding Equipment 4, 20, 35, 44, 47
Nozzle Cleaners, MIG-Welding 13, 35, 44, 75
Plasma-Arc Cutting & Welding Equipment 18, 24, 44, 47, 72
Power Supply, See Specific Types
Remote Controls, Arc-Welding 13, 16, 18, 35, 44, 47, 69
Respirators 27, 44, 48, 52
Robot Welders 35, 68, 82
Rods, Gas-Welding 6, 7, 7A, 10, 22, 26, 44, 46, 53A, 56, 71, 75, 76
Schools & Training Courses 4, 8, 26, 33, 34, 62
Soldering Alloys & Fluxes, Soft 5, 7, 8, 22, 25, 26, 42, 54, 71, 75
Spot Welders, Portable 9, 36, 43, 47, 59, 68A
Spot Welders, Stationary 9, 28, 36, 59

Spot-Welding Guns 9, 35, 36, 44, 59
Stainless-Steel Back-Shield Paste 63
Temperature Indicators 53, 66
TIG-Welding Equipment 4, 17, 19, 24, 26, 35, 44, 73, 75, 81
Tip Cleaners 4, 23, 33, 44, 45A, 75,
Torch Lighters 4, 22, 23, 33, 44, 75
Torches, Cutting, Underwater 4, 9, 13
Welding & Cutting Kits 4, 23, 26, 33, 38, 40, 44, 60, 62, 75, 77, 83
Wire Feeders, Semi-Automatic 4, 10, 18, 24, 35, 44, 47, 64, 69

Metric Conversions

METRIC CUSTOMARY-UNIT EQUIVALENTS

Multiply:	by:		to get:	Multiply:	by:		to get:
LINEAR							
inches	X 25.4	=	millimeters(mm)	X	0.03937	=	inches
miles	X 1.6093	=	kilometers (km)	X	0.6214	=	miles
inches	X 2.54	=	centimeters (cm)	X	0.3937	=	inches
AREA							
inches^2	X 645.16	=	millimeters^2(mm^2)	X	0.00155	=	inches^2
inches^2	X 6.452	=	centimeters^2(cm^2)	X	0.155	=	inches^2
VOLUME							
quarts	X 0.94635	=	liters (l)	X	1.0567	=	quarts
fluid oz	X 29.57	=	milliliters (ml)	X	0.03381	=	fluid oz
MASS							
pounds (av)	X 0.4536	=	kilograms (kg)	X	2.2046	=	pounds (av)
tons (2000 lb)	X 907.18	=	kilograms (kg)	X	0.001102	=	tons (2000 lb)
tons (2000 lb)	X 0.90718	=	metric tons (t)	X	1.1023	=	tons (2000 lb)
FORCE							
pounds—f(av)	X 4.448	=	newtons (N)	X	0.2248	=	pounds—f(av)
kilograms—f	X 9.807	=	newtons (N)	X	0.10197	=	kilograms—f

TEMPERATURE

Degrees Celsius (C) = 0.556 (F - 32)   Degree Fahrenheit (F) = (1.8C) + 32

ENERGY OR WORK

foot-pounds	X 1.3558	=	joules (J)	X	0.7376	=	foot-pounds

FUEL ECONOMY & FUEL CONSUMPTION

miles/gal	X 0.42514	=	kilometers/liter(km/l)	X	2.3522	=	miles/gal

Note:
235.2/(mi/gal) = liters/100km
235.2/(liters/100km) = mi/gal

PRESSURE OR STRESS

inches Hg (60F)	X 3.377	=	kilopascals (kPa)	X	0.2961	=	inches Hg
pounds/sq in.	X 6.895	=	kilopascals (kPa)	X	0.145	=	pounds/sq in
pounds/sq ft	X 47.88	=	pascals (Pa)	X	0.02088	=	pounds/sq ft

POWER

horsepower	X 0.746	=	kilowatts (kW)	X	1.34	=	horsepower

TORQUE

pound-inches	X 0.11298	=	newton-meters (N-m)	X	8.851	=	pound-inches
pound-feet	X 1.3558	=	newton-meters (N-m)	X	0.7376	=	pound-feet
pound-inches	X 0.0115	=	kilogram-meters (Kg-M)	X	87	=	pound-inches
pound-feet	X 0.138	=	kilogram-meters (Kg-M)	X	7.25	=	pound-feet

VELOCITY

miles/hour	X 1.6093	=	kilometers/hour(km/h)	X	0.6214	=	miles/hour

Index

Basic Welding Symbols and Their Location Significance

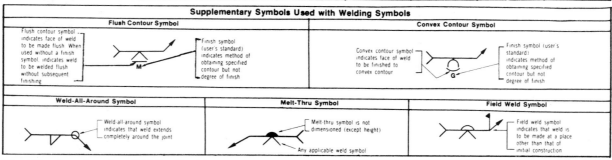

Chart Courtesy American Welding Society